ELECTRONICS MANUFACTURING: Devices and Processes

by Barry M. Lunt, Ph.D.

School of Technology
Brigham Young University

First edition: Prentice Hall, Upper Saddle River, New Jersey, 2004
Second edition, with chapters 1-12 updated and chapters 13-19 removed: Brigham Young University Press, 2014

ISBN 978-0-8425-2887-0

Dedication

To my loving and tremendously supportive wife, Cathy.
To my awesome children, Greg, Shanda, Cindy, Doug and Brad.
To my beloved children-in-law, Sarah, David and Laken.
To the most amazing grandchildren on this planet.
And to everyone who has greatly supported me in this undertaking.

Preface

Approximately ¼ of all the manufacturing in the world today, in US $, is electronics, from the largest datacenters to our increasingly popular smart phones. These devices represent some of the most astounding advances in manufacturing in the history of the world. Even if we could only make a few of these per year, it would be astounding. But we make billions of them every year, and the cost continues to fall as the performance improves, something that does not happen in any other field of manufacturing. It is the purpose of this book to help understand the manufacturing processes that have made and continue to make this possible.

This book does not cover every electronic device in use today. Specifically, it does not cover the manufacturing of fuses, relays, solenoids, motors, wire, cables, optical fiber, thermistors, or other highly specialized devices. But it does cover the big three: capacitors, resistors, and transistors, and the modern integrated circuit (IC), which includes all three of these elements. It also covers connectors, printed wiring boards (PWBs), backplanes, and the process of assembling PWBs. These devices and processes are the core of today's electronics industry, as they have been for decades, and it appears they will continue to be for a long time to come.

In this history of the world, no product has ever become so widely adopted and made in such large numbers as the transistor. Originally invented in 1947, the IC industry today makes over 200 billion of these tiny devices for every person on the planet, every year. That's over 1.4 septillion (1.4×10^{21}) – every year! Resistors and capacitors are also made in very large numbers, but are still many orders of magnitude fewer than transistors.

It is this author's hope that the world can continue to enjoy the benefits of these amazing devices, and that we can continue to turn out manufacturing engineers who can continue to contribute to the many advances being made in this industry every year.

CONTENTS

CHAPTER 1: INTRODUCTION

CHAPTER OVERVIEW

This chapter covers the following concepts:

- The changes wrought in society by advances in the field of electronics
- A brief history of the development of our understanding of electricity and its applications in electronics
- The dangers we should be aware of when working with electricity
- The organization and general approach of the text

1.1: ADVANCES IN ELECTRONICS

For thousands of years prior to the mid-1800s, people traveled by two main methods: they used their own power, or they used the power of domesticated animals. With few exceptions, information traveled in much the same manner. A person had to carry the information in order for it to reach its destination. Travel was facilitated as the automobile and airplane were developed. Distances that formerly took days, weeks, or even months to travel now could be completed in less than a day. Through major advances in manufacturing technology, automobiles became accessible to the general public. The changes wrought on society by these advances in transportation technology have been profound.

As significant as these changes in transportation technology have been, they pale in comparison to those wrought in the field of information transmission and processing. These constitute the modern fields of electronic communication and information processing. Table 1.1 compares these changes over the past fifty years.

Indeed, as demonstrated in Table 1.2, if transportation technology had been able to improve at the same rate, today's automobile would be vastly different.

As a result of these changes in the fields of communication and information processing, the modern world is dramatically different compared what it was only a few decades ago. Prior to 1950, the main electronic products available to the modern consumer included telephones, televisions, radios, phonographs, and amplifiers. Today a list of electronics products available to the public would occupy this entire page.

(IP)	Transportation		Communication		Information Processing	
	1950	**2015**	**1950**	**2015**	**1950**	**2015**
Top Speed	80 mph	550 mph	~1,000 char/ sec	1e10 char/sec	5000 instr/sec	1e12 instr/sec
Purchase Cost	$2000 $10/100 miles	$15,000 $50-100/100 miles	$1,000/MByte/ sec	$20	$200M / MIPS	$10/MIPS
Efficiency	15 mpg	30 mpg	1 MB/sec/KW	.0005	.025 MIPS/Watt	10M
Weight	4,000 lbs.	3,000 lbs.	1 ton/MB/sec	¼ lb.	50 KIPS/ton	50 GIPS/ton
Life	80,000 miles	150,000 miles	30 MB * Yrs	400,000 MB * Yrs.	25 KIPS * yrs	10,000 MIPS * yrs

Table 1.1 A comparison between advances in transportation, communications, and information processing.

		2015	
	1950	Using Communication Comparison	Using IP Comparison
Speed	80 mph	400,000,000 mph / 10,000 mph (with wind resistance)	3.6 Gmph / 37,000 mph (with wind resistance)
Operating Cost	$10/100 miles	$0.008/100 miles ($8,000 / 100 M miles)	1 n¢ / 100 mi. (10¢ / 100 M miles)
Efficiency	15 mpg	27,000 mpg	7.20 Gmpg
Weight	4000 lbs.	0.75 lbs.	.002 lbs (~0.032 oz)
Life	80,000 miles	500 M miles	12.0 G miles
Purchase Cost	$2000	$0.90	$0.25

Table 1.2 What today's automobile would be like if it had been improved as often and as quickly as communications and computers were.

Two major breakthroughs in the field of electronics account for this phenomenal change: the transistor and the integrated circuit. But to adequately understand how these came about, a bit of history is in order.

1.2: HISTORY OF ELECTRICITY AND ELECTRONICS

As far back as 300 BC, the Greeks observed and studied the effects of static electricity. By rubbing a piece of amber (elektros is Greek for amber) with one material, a charge could be created. By rubbing it with another material, the second material could be made to attract the first material. The Greeks explained this using two types of electric fluid of opposite natures.

Until the late eighteenth century, very little progress was made in the field of electricity. However, starting in the nineteenth century, several experimenters put together key bits of information regarding the nature of electricity.

Alessandro Volta produced the voltaic pile (battery), the first continuous form of electricity, (as opposed to static electricity).

Research at the University of Leyden in the Netherlands showed that electricity (then considered an electrical fluid) could be stored in specially made jars. These jars became known as Leyden jars (see Figure 1.1). Leyden jars were actually the first capacitors.

Figure 1.1. Leyden jars, the first capacitors. *Courtesy of PV Scientific Instruments, http://www.arcsandparks.com.*

Benjamin Franklin showed that the phenomenon we know as lightning was nothing more than a magnificent display of static electricity; he also gave the two types of electrical fluid their present designations of negative and positive electricity.

In 1819 Hans Christian Oersted showed that a magnetic field existed around a wire that was conducting electricity.

Michael Faraday followed in 1831 when he showed that electricity was produced in a wire in the presence of a changing magnetic field.

He then invented a mechanical device capable of producing electricity—the world's first electrical generator. Soon there appeared practical applications of this emerging understanding; the single-wire telegraph in the 1860s, the telephone in the 1880s, and electrical motors of many types in the mid-to-late 1800s. All of these inventions proved a great boon to civilization when put to their proper uses.

As significant as these developments were, they were only the foundation of more dramatic upcoming changes. In the late 1800s, James Clerk Maxwell laid a solid foundation for understanding the propagation of radio waves. This was followed by careful and successful experiments by Heinrich Hertz, who established that wireless electrical signals could be detected over short distances. Continuing in this work, Guglielmo Marconi showed in 1901 that wireless communication (radio) could traverse the entire Atlantic Ocean. In 1904 Ambrose Fleming expanded on the Edison effect, first noted by Thomas Edison. Edison had observed that an electrical current would only flow from the cathode to the anode, and not vice versa. Fleming's thermionic valve (see Figure 1.2) became the first vacuum tube rectifier, and was a great aid in radio detection.

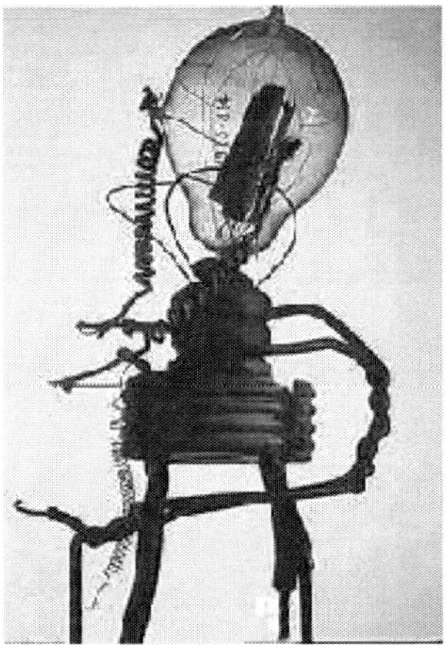

Figure 1.2 Fleming's Thermionic Valve. *Courtesy of the Science Museum, London, England.*

In 1906 Lee de Forest further developed the thermionic valve into the audion (see Figure 1.3), the first electronic amplifying device.

Figure 1.3 Lee de Forest holding his two-element audion. *From Rosenblatt, Alfred. June, 2000. The Electric Century. IEEE Spectrum. © IEEE*

This was significant for two key reasons: first because it could be used as a switch or relay, and second because it was the first way ever invented to amplify signals. It also was a much faster switch, operating over one thousand times faster than relays. Its ability to amplify finally gave the world a way to take faint signals, such as those from microphones or antennas, and amplify them for public address (PA) systems, radio, or recording. Soon after came AM and FM radio, "portable" radios, television (about 1930), and long-distance wireless voice communication.

The next major event in the history of electronics was the invention of the transistor, first successfully demonstrated in December of 1947 (see Figure 1.4) at Bell Laboratories by John Bardeen, William Shockley, and Walter Brattain, who later won the Nobel Prize for their invention.

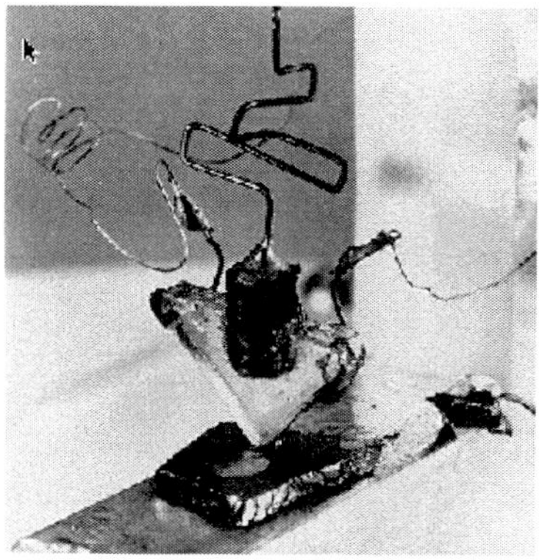

Figure 1.4 The first transistor, Circa 1947. *Courtesy of Bell Labs, Lucent Technologies.*

The transistor could do most things the vacuum tube could do, but it had seven big advantages, each of which was very significant. These advantages consisted of the following:

1. They were smaller.
2. They weighed less.
3. They used less power.
4. They were more reliable.
5. They were more rugged.
6. They soon became less expensive.
7. They turned on instantly (required no warm-up).

With the invention of the transistor began the development of powerful computers, advanced communication devices, and truly portable battery-powered devices. But this was only the start.

Only twelve years later (1959) came the invention of the integrated circuit, (see Figure 1.5) developed independently by Jack Kilby at Texas Instruments and Robert Noyce at Fairchild Semiconductor.

Figure 1.5 The first integrated circuit. *From Rosenblatt, Alfred. June,2000. The Electric Century. IEEE Spectrum. ©IEEE.*

The integrated circuit allowed for the four most common electronic circuit elements, (transistors, resistors, diodes, and capacitors) to be fabricated on a common piece of semiconductor material. Its significance was that it once again brought many advantages; in fact, all those same advantages brought by the transistor:

1. It was smaller.
2. It weighed less.
3. It used less power.
4. It was more reliable.
5. It soon became less expensive.

Early integrated circuit clips had only about ten transistors per chip, but by the late 1960s, IC's were being made with over 200 transistors per chip. This increase in the density and number of transistors has increased at the rate of a doubling every eighteen months, a trend which Gordon Moore of Intel commented upon, and that came to be called "Moore's Law." How much longer "Moore's Law" can hold true is now uncertain, but the result of more than fifty years of this trend has today given us IC's with more than 100 billion transistors in an area smaller than 1 in^2, and has made the computer ubiquitous. It is primarily the IC which has made possible today's explosion of electronic products, many of them inexpensive and very helpful. It is also primarily the advances in IC technology that enable the electronics industry to continue to deliver improved products and reduced prices with each iteration.

1.3: SAFETY

The next chapter will introduce the broad topic of physical design of electronic products. Before we leave this introductory chapter, it would be wise to spend some time addressing the issue of safety. It is nearly inevitable that those who work with electricity will come into contact with it, as undesirable as this is. Because we are very likely to come into contact with electricity, we should first be armed with the knowledge of how to minimize our chances of being hurt by it.

1.3.1: Indirect Dangers

Electricity can hurt us either directly or indirectly. Indirect ways of being hurt by electricity include accidental falls, accidentally dropping something, fire, and explosion.

If we are working on a ladder or an elevated platform, it is possible that accidentally coming into contact with electricity will cause us to spring back involuntarily, either from the surprise, or from the physical shock of the electricity. This sudden movement can cause serious injury.

If we are holding something heavy or fragile, it is possible that inadvertently coming into contact with electricity will cause us to drop the object we are holding. A heavy object could injure us or someone near us; a fragile object could break. If the object breaks and a dangerous substance (flammable or explosive substances, toxic gases) is released, a further danger results.

A sudden electrical short can easily start a fire in flammable material. If working with both electricity and combustible materials, great care must be taken to keep the two separated. Proper safety procedures require that all electrical devices be heavily shielded in areas where highly flammable materials are in the open. If working with flammable liquids or dust, the area should be well ventilated, and free from open flames or sparks.

The final indirect danger (explosion) is very similar to fire, because an explosion is the very rapid combustion of a volatile material. Explosive environments include grain elevators (these have been known to explode from a small electrostatic discharge, or ESD), wood sanding areas, almost any area with high dust levels, areas near bottled gases (propane, butane), filling stations, or anywhere that there can exist a concentration of volatile liquid vapors. Great precautions must be taken in these environments to prevent ESD or other types of open electrical sparks, such as open brush motors or relays.

1.3.2: Direct Dangers

The direct danger of electricity is that even without any of the indirect dangers present, electricity can kill. In spite of many precautions and the increased level of public awareness, each year many people die from electric shock. In almost every case, the tragedy could have been avoided by following some basic safety rules. Before we get to the rules, let's study the theory behind how these rules protect us.

Table 1.3 shows us the various levels of physical sensation that result from exposing the body to electricity.

Current (milliAmps)	Physical Sensation
< 3	Not sensed
3 - 10	Mild sensation
10 - 20	Painful
20 - 30	Cannot let go
30 - 40	Muscular paralysis
40 - 100	Severe shock; can be fatal
100 - 200	Usually fatal
200 - 1000	Severe burns; breathing stops; often fatal
> 1000	Dismemberment

Table 1.3 The physical effects of electric currents.

As can be seen, the part of electricity that is dangerous is the current flowing through the body, and not the voltage. Then why do signs say, "Danger: High Voltage"?

This brings us to what we can refer to as the Ohm's Law of safety. Ohm's Law says that the current that flows is proportional to the voltage. Clearly, the higher the voltage to which we may be exposed, the greater the current can be, depending on the resistance of the path. Low voltages are generally considered safe; <50 Volts is usually not enough, except in very extreme circumstances, to cause death by electrocution (direct danger). This is why handling jumper cables on 12-Volt car batteries, or putting a small 9-Volt battery on your tongue, is not considered dangerous; the voltage is simply too low to cause a serious amount of current to pass through the body.

The other side of the Ohm's Law of safety is where the safety rules come in. Ohm's Law also says that the current that flows is inversely proportional to the resistance. This means that if the resistance through our body is sufficiently high, we cannot be directly harmed by electricity. The extreme example of this is often referred to as the "bird on a wire" analogy. Though a bird sitting on a high-voltage wire is directly exposed to a very high voltage, the bird's body is not a path for the electricity because it is floating in the air, and for this reason no current flows. But if the bird were an ostrich, and its head came into contact with the same high voltage, a great deal of current would flow because the feet of the ostrich are grounded.

We generally cannot control the voltages we may be accidentally exposed to. For instance, in our homes, we have essentially no control over the fact that all potential sources of electric shock are 120 Volts, which is well into the potentially lethal range. So the only part of the Ohm's Law of safety that we have control over is our resistance. Common sense safety rules specify never to work on 120-Volt electrical circuits in our bare feet, or in standing water, even with shoes on. Standing well-grounded, as we would be in these cases, would make our bodies into low-resistance paths for current—a deadly situation. If you must work on 120-Volt circuits, first make sure the circuit is not energized, then make sure you are not grounded (meaning you are wearing dry shoes with no holes, and you are not lying on the floor). That way, if the circuit is accidentally energized, your shock will be minimized.

Up until about twenty years ago, we often dealt with lethal combinations of electricity and water in our bathrooms, kitchens, and outdoor outlets. In these rooms or outdoors, people often have no shoes on, and electrical outlets are available. If a faulty electric appliance is used in such a situation, the accidental contact with electricity can be lethal. Also, if an appliance is not properly grounded, the results could be deadly. This has led to requirements for Ground Fault Interrupt (GFI) outlets in any locations where the user could be standing on the ground (outdoors) or barefoot on a wet floor (bathrooms, even kitchens). These have undoubtedly saved thousands of lives over the years, with many thousands more to come. GFI's work by monitoring the current flowing out of the outlet, and comparing it to the current which flows back into the outlet. Ideally, the two currents should be identical; if they are not, a short (ground fault) is the reason, and the GFI senses this and trips the interrupt in as little as 1 millisecond, before a lethal exposure can result. GFIs are so effective that for many years, they have been required on hair dryers, so that even if the bathroom does not have a GFI outlet (many older homes do not), the user cannot die if the hair dryer falls while they are using it in the tub. (Don't laugh; this has happened many times!)

Another good safety rule is that if you must work with an unsafe voltage (>50 Volts), use only one hand. This is true because if only one hand is being used, it is not possible for one hand to be grounded and the other to come into contact with the voltage. This combination would be particularly dangerous because the most common physical cause of electrocution is stoppage of the heart, or fibrillation. With the hands forming the circuit, the path of the current is through the heart. When the heart receives conflicting electrical signals, such as those arriving from the brain and those from the electrical circuit, it often (about 20% of the time, depending on the phase of the heartbeat) goes into fibrillation, a condition in which the heart simply quivers. Fibrillating hearts pump no blood, and unless normal beating is restored within minutes, death will result. If you're using only one hand (with the other in a pocket, or somewhere it cannot come into contact with a ground) and your free hand contacts the voltage, the only path is through your arm and to your feet, which avoids the path through the heart. Additionally, if you are wearing shoes as you should be, your resistance will be high, and only a very low current will result.

In conclusion, remember that electricity can be very deadly. Hundreds of people die from it every year. Use all necessary precautions when working with anything above 50 Volts, and remember the dangers of indirect injury with even lower voltages. We do not need more people to underline the statistics that already clearly show the danger.

1.4: INTENT AND APPROACH OF THIS TEXT

This text is written on the assumption that the reader has already had a class in classical physics, and a class in common manufacturing methods and processes. It also assumes basic familiarity with devices such as resistors, capacitors, and inductors. This text will introduce manufacturing engineers to the most common electronic devices (resistors, capacitors, inductors, transistors and integrated circuits) and the processes used to manufacture these devices.

CHAPTER SUMMARY

The changes brought about by the past century's advances in electricity and electronics are so significant that there is no other period in history to even compare with it. There are myriad products that exist today only because the transistor and the integrated circuit have made it possible to make them, and this trend shows no signs of letting up.

In order to appreciate where we are today in this rapid advance, it is helpful to review the history of this field. Countless individuals have made very significant contributions and discoveries, and all have built on the base of the knowledge created by predecessors. This pattern of constant progress is very likely to continue.

We have welcomed electricity into our homes, automobiles, businesses and industries, but it has not been without a cost. Thousands of people are injured or killed by electricity every year, so it is urgent that those working with electricity understand the dangers involved. These dangers are both direct and indirect.

FOCUS QUESTIONS

1. How many orders of magnitude improvement are represented in the speed of information processing over the last 50 years? How does this compare to the improvement represented in the weight and purchase cost of communication equipment over the last 50 years?
2. How long have we been aware of the existence of (static) electricity?
3. What was the practical significance of the invention of the audion (later known as a vacuum tube)?
4. What were the advantages of the transistor over the vacuum tube? How do these advantages compare to the advantages of the integrated circuit over the

transistor?

5. List three ways electricity can indirectly be a danger to us.
6. What is the main direct danger with electricity? What can you do to reduce the risk of this danger?
7. How does a GFI prevent lethal shocks?

FOR ADDITIONAL READING

Blackwell, Glenn R., “The Electronic Packaging Handbook”, Chapter 16, CRC Press, Boca Raton, FL, 1999.

Boylestad, Robert L., “Introductory Circuit Analysis”, 8th edition; section 1.2.

Dummer, G. W. A., “Electronic inventions and discoveries : electronics from its earliest beginnings to the present day”, Institute of Physic publishing, Philadelphia Pa, 1997.

Electronics - editors, “An Age of innovation : the world of electronics 1930-2000", McGraw-Hill, New York, NY, 1981.

Lunt, Barry M., “Safety with Electricity”, The Technology Interface, vol 3 #3, Fall 1999.

Matisoff, Bernard S., “Handbook of Electronics Packaging Design and Engineering”, Chapter 15, Van Nostrand Reinhold, New York, NY, 1990.

Wolf, Stanley, and R.F.M. Smith, “Student Reference Manual for Electronic Instrumentation Laboratories”, Chapter 1, Prentice-Hall, Englewood Cliffs, NJ, 1990.

CHAPTER 2: ELECTRONIC COMPONENTS AND PACKAGING LEVELS

CHAPTER OVERVIEW

This chapter covers the following concepts:

- There are several packaging levels; the first four of these are covered in this text
- There are tens of thousands of choices when it comes to choose an IC
- IC packages have many purposes; no single package type is always best
- There are several types of IC substrate materials that can be used for making ICs

2.1: INTRODUCTION

There are several levels of packaging (at least seven) and several types of packaging used in current electronic products. The first section of this chapter gives a brief overview of these packaging levels in electronics products. The next seven sections cover some of the physical characteristics of level 1 electronic devices. The final section is a conclusion of the material covered in this chapter.

2.2: PACKAGING LEVELS

In electronic products, as in many other types of products, the product as a whole is made up of many smaller parts assembled together to form the whole. The hierarchy of packaging levels for electronic products in general is depicted in Figure 2.1, along with one of the names commonly used for each level. Formally, these packaging levels have also been given numerical designations: level 0 is the chip, also called the die or IC; level 1 is the module or IC package; level 2 is the card, PWB, or sometimes (confusingly) the board; level 2 is the board, motherboard, or backplane.

There are packaging levels above those depicted in Figure 2.1: level 4 is the rack, which contains multiple backplanes; level 5 is the gate, which contains multiple racks; and level 6 is the frame, which can contain multiple gates. These latter three packaging levels are never seen in consumer equipment and so will generally not be familiar to the reader; however, large products such as servers, supercomputers, and telecommunications switching equipment sometimes use these packaging levels.

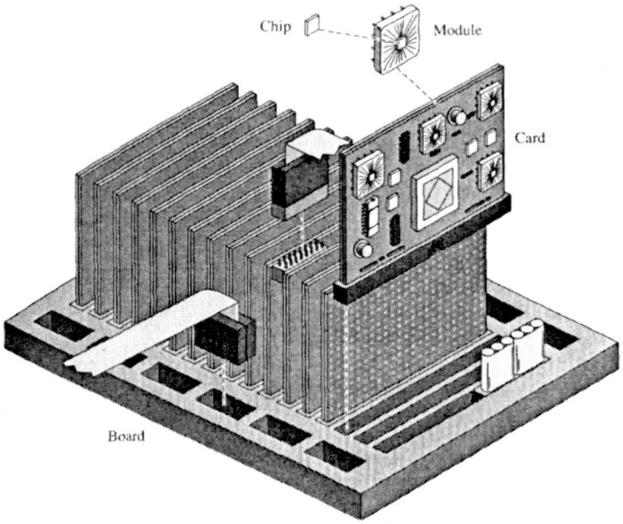

Figure 2.1 Four electronics packaging levels. *Seraphim, Donald, Ronald C. Laksy, and Che-Yu Li. 1989 Principles of Electronic Packaging. New York: The McGraw-Hill Companies. Reproduced by permission of the McGraw-Hill Companies.*

2.3: INTEGRATED CIRCUITS

ICs are, by far, the most heavily used of all level 1 devices in electronics. They come in many kinds, far more than could be referenced in a single chapter. This section gives an idea of the choices that exist.

2.3.1: Functional Blocks

In designing electronic circuits, ICs are usually chosen first for the function or functions they perform. The field of IC functions can be broken into two main categories: analog and digital. Examples of analog functional blocks include amplifiers, comparators, phase-locked loops, timers, voltage and current sources, and voltage references. Examples of digital functional blocks include the basic logic primitives (AND/NAND, OR/NOR, NOT), advanced basic functions (multiplexors, counters, latches, etc.), and system functions, such as microprocessors, display drivers, memory, etc. There is an immense array of most of these functions, both analog and digital, in a wider range of functional and performance specifications. For further details on the "off-the-shelf" technology building blocks that are available, the reader is encouraged to consult the data books on websites of the companies that produce the circuits. A brief summary of the range of choices available is found in Tables 2.1, 2.2, and 2.3.

Functional Block	Approximate Number of Choices Available
Amplifiers, Comparators	10,000
Clocks, Synthesizers	3000
Consumer Circuits	10,000
Power Converters, Power Control	9200
Optoelectronic	3100
Telecommunications	8100
Voltage Regulators	8500
Other	5200

Table 2.1 Table of Analog Circuits

Functional Block	Approximate Number of Choices Available
Buffers	4200
Bus-oriented	7400
Counters	3900
Delay lines	5200
Gates	6400
Flip Flops	4000
Interface	11,500
Latches	3000
Memory (DRAM, SRAM, ROM, etc.)	84,000
Microcontrollers, microprocessors, DSPs, System components	19,200
Programmable Logic Devices	6100
Standard Cell, Gate Array, Mixed Array	4300
Other	14,500

Table 2.2 Table of Digital Choices

Functional Block	Approximate Number of Choices Available
Mixed Array	415
A-D Converters	4800
D-A Converters	4800

Table 2.3 Table of Mixed Circuits

2.3.2: IC Packages

Packaging levels were introduced previously in section 2.2. IC packages are level 1; this is the packaging level that will be focused on in this chapter. Level 0 (the actual die or chip) will be covered in chapters 7, 8 and 9.

After deciding the IC functions needed, the designer must next decide the package type that the IC will come in. Many package types have been developed, each with its own set of characteristics. IC packages have been designed with several purposes in mind. These purposes are discussed below.

2.3.2.1: Cost: Low

Generally, the IC package should be as inexpensive as possible. Several requirements are necessary for this to happen:

1. It must be made from inexpensive materials
2. It must be mass manufacturable, meaning it can be assembled using automated equipment, and does not require too many manufacturing processes.

Modern IC packages typically range in cost from two cents to ten dollars, with some very high-end packages costing much more.

2.3.2.2: Environment: Sealing Against It

The goal is for the package to prevent moisture or other gaseous vapors from penetrating through to the chip inside. Although, when IC's are made, steps are taken to prevent them from being sensitive to such vapors (specifically, they are "passivated"), it is still highly desirable that the IC package provide an additional degree of protection from the permeation of these vapors. This means that IC package materials should provide an hermetic (air-tight) seal. The rate at which moisture permeates modern IC packages ranges from a few seconds to many centuries, depending primarily on the type of packaging material (see Chapter 9).

2.3.2.3: Heat: Removal

ICs always produce heat when in operation, so the goal here is for the package to conduct the heat away from the chip and convect it to the surrounding air. This means that package materials should be good conductors of heat. Materials used in today's IC packages have a thermal conductivity from .2 to 6000 W/mK.

2.3.2.4: I/Os: Many

As shown in Figure 2.2, the trend for decades was for the number of I/Os to keep going up. This trend has basically reached a practical maximum, even though more I/Os would be desirable. Ideally, IC packages need to be capable of providing as many I/Os as the IC chip requires. Modern ICs typically have as few as 3 to as many as 3,000 I/Os. While more I/Os would be useful in many instances, it seems we have reached a practical limit.

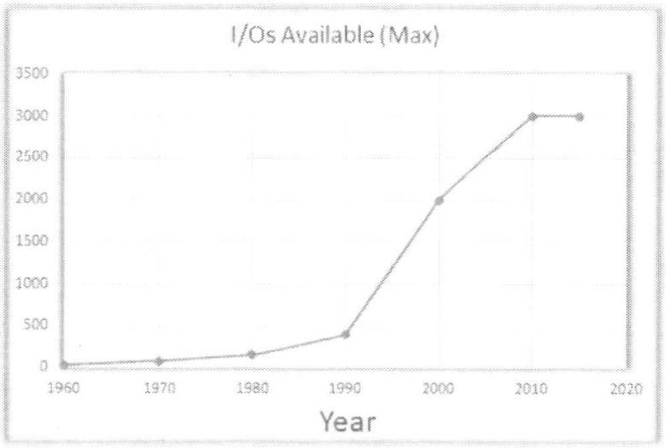

Figure 2.2 Module I/Os as a function of time.

2.3.2.5: Level 2 Assembly: Easy

Level 2 refers to the printed wiring boards (PWBs) to which most IC packages are attached. Because this attaching process is usually automated, it is very desirable that the IC package not increase the difficulty of this process. The main factor in IC packages that makes this process more difficult is the lead pitch, which is the distance between the centers of adjacent leads, as shown in Figure 2.3. The finer the pitch (the closer the leads,) the greater the difficulty involved in level 2 assembly. It can be seen, however, that this requirement directly conflicts with the preceding requirement of many I/Os; the closer together the leads are (the finer the pitch), the more of them we can fit around the periphery of the package.

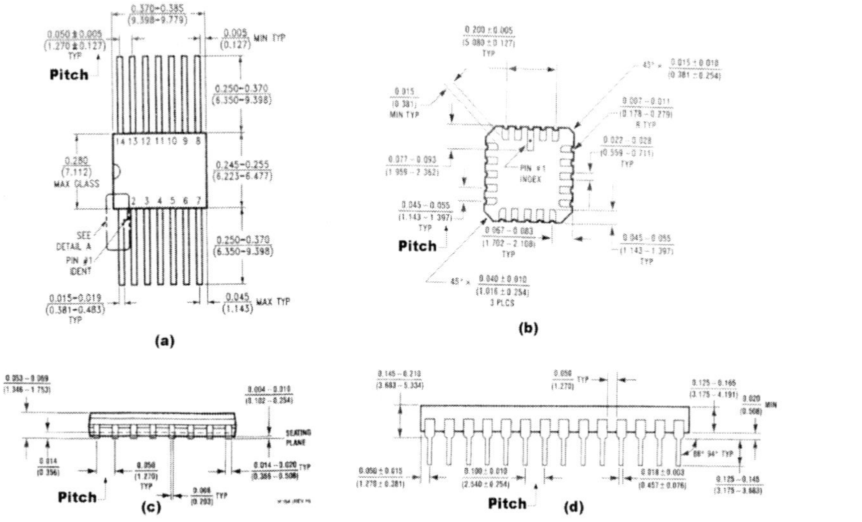

Figure 2.3 Four examples of lead pitch. a) flatpack; b) leadless ceramic chip carrier (LCCC); c) small outline IC (SOIC); d) dual in-line package (DIP). *Courtesy of National Semiconductor.*

2.3.2.6: Parasitic Properties: None

The process of getting power and signals in and out of an IC, without changing them in an adverse way, is often a daunting challenge. All IC packages add to this challenge, to one degree or another, in the form of parasitic resistance, capacitance or inductance. Therefore, the goal is to have IC packages with the smallest possible parasitic properties. Modern packages can have parasitic resistance from 1mΩ to 1Ω, parasitic capacitance from .1 pF to 50 pF, and parasitic inductance from .1 nH to 100 nH, for each I/O. Figure 2.4 gives an example of a model of these parasitics.

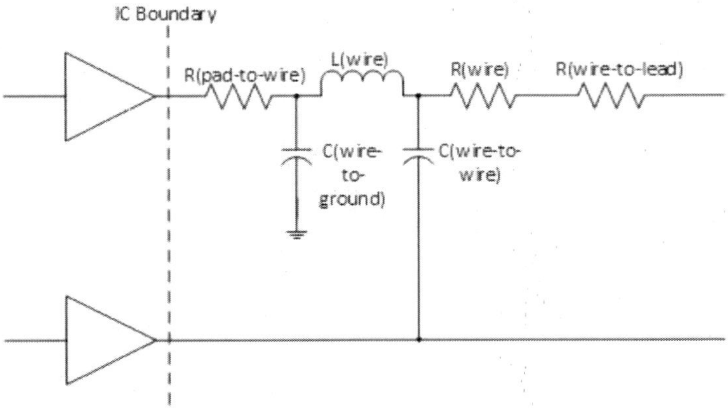

Figure 2.4 A model of the parasitic devices on package leads. A lead is not always a simple wire, especially at very high frequencies

2.3.2.7: Physical Protection: Heavy Duty

IC chips are much more rugged than their predecessor, the vacuum tube. However, without a package they are not rugged enough to withstand the rigors of ordinary shock and vibration to which they are routinely subjected, much less the rigors of high shock and vibration applications such as automobiles, missiles, and hockey pucks. It is therefore desirable that the IC package provide as much protection as possible in this area. Today's IC packages protect the IC against shock and vibration from 100 g's up to 20,000 g's.

2.3.2.8: Reliability: High

Three of the main factors of high-reliability IC packages have already been addressed under the headings Environment, Heat, and Physical Protection. There is one additional factor, and that is power cycling. Each time an IC turns on, it heats up; when it turns off, it cools down; this is one power cycle. The main problem this causes in terms of reliability comes from the fact that nearly all solid materials expand when heated and contract when cooled. This would not be a problem in IC packages if all materials expanded and contracted at the same rate, but they do not. Each power cycle puts these dissimilar materials, with their dissimilar coefficients of thermal expansion (CTE), through one cycle of stress and strain. The goal in this area of high reliability is to have IC packages that can withstand many power cycles, in addition to the environment, heat, and physical protection requirements mentioned earlier. Modern IC packages can withstand from a few hundred cycles to tens of thousands of power cycles, depending heavily on the difference between the operating temperature and the non-operating temperature of the IC.

2.3.2.9: Size: Small

One of the big advantages ICs brought was the reduction in the physical size of the circuit. Size reduction continues to be one of their biggest advantages, so it only follows that we would also want the IC package to be small. The lower limit on the size of the IC package is the size of the IC chip itself. Recently, many package types have evolved to help reduce the size of the IC package; some of these are shown in Figure 2.5. Figure 2.6 shows some of the other IC packages that have evolved to meet other needs.

Modern IC packages range from only slightly larger than the IC chip (1.1x on either dimension), to many times larger (up to 6x on either dimension).

Figure 2.5: Tiny IC chip-scale packages (CSPs) on the face of a penny. *Advanced Packaging (September/October 1998): 50. Published by permission.*

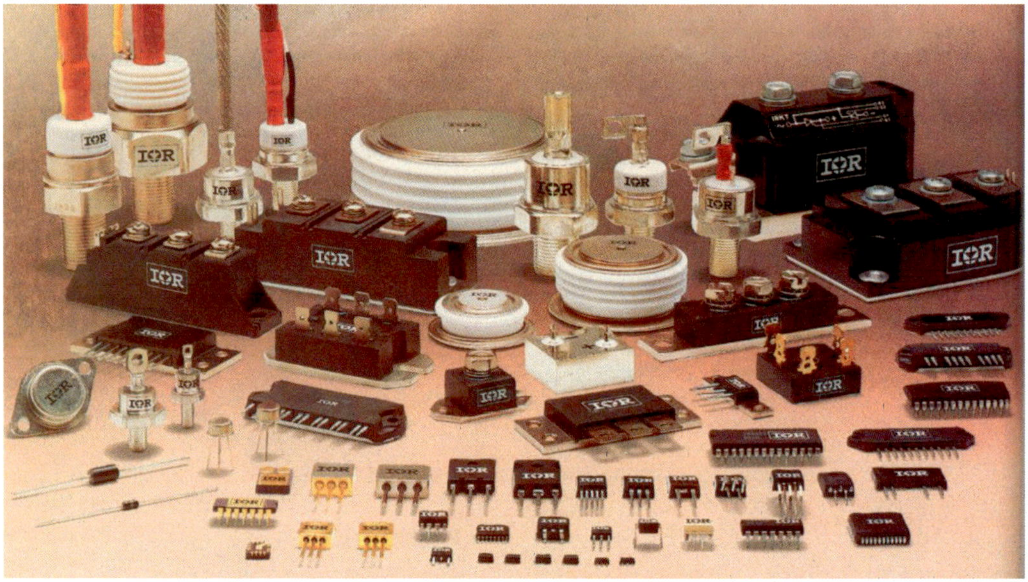

Figure 2.6 Variety of IC packages. *Electronic Design (November 11, 1993): 24. Courtesy of International Rectifier.*

2.3.2.10: Testability: High

After an IC chip is fully assembled in its package, it must be tested for proper operation. It is desirable that the test process not add significantly to the cost of the IC, which means that testing must be automated. It is therefore desirable that the IC package not impose additional challenges to automation of test. All popular IC packages are very successful in meeting this goal.

2.3.2.11: Weight: Low

In many ways, this ties in directly with the size of the package, which was discussed above. However, this area also includes the thickness of the package (it should be as thin as possible), and the density of the materials from which the package is made (the lower the better). Weight is not a major factor in many applications, but in portable products and products for airplanes, rockets, and satellites, weight is critical.

2.3.2.12: Package Options

Today's designers have many packages to choose from, made from several materials, and with a wide variety of characteristics. Some of these are shown in Figure 2.6. A comparison of many IC packages is found in Table 2.4; each package is compared for its performance in nine of the preceding eleven areas. They are not compared for physical protection nor for testability, because all of the popular packages perform very well in these areas. Readers are encouraged to become familiar with this table and to use this information in choosing the right IC package type for their applications.

A few generalizations based on Table 2.4 are in order here. It can be noted from the columns, "Environment," "Heat," and Reliability" that the package performance in these areas is dominated by the packaging materials: ceramic is better than plastic for sealing against the environment, for removal of heat, and for power-cycling reliability. However, it should also be noted that for identical size packages, ceramic packages are heavier than plastic ones. The details behind these observations will be covered in chapter 10.

(1=Very Good; 5=OK)

Type	Level 2 Assembly	Cost	Environment	Heat	Number of I/0s	Parasitics	Reliability	Size	Weight
DIP (Plastic)	2	1	3	4	5	5	4	5	4
PPGA	2	2	3	4	2	3	4	3	4
CPGA	2	4	1	1	2	3	1	3	5
SOIC (Plastic J-lead)	2	1	3	4	3	3	4	3	2
SOIC (Plastic Gull-wing)	2	1	3	4	3	3	4	3	2
PLCC	2	2	3	4	3	3	4	3	2
CLCC	2	4	1	1	3	3	1	3	3
Flatpack (Plastic)	2	1	3	4	3	3	4	3	2
BGA (Plastic)	3	2	3	4	1	2	4	2	2
BGA (Ceramic)	3	4	1	1	1	2	1	2	3
Fine-pitch peripheral leads(Ceramic)	5	3	1	1	2	3	1	3	2
Fine-pitch peripheral leads(Plastic)	5	3	3	4	2	3	4	3	2
CSP's	2	2	C=1, P=3	C=1, P=4	3	1	C=1, P=4	1	1

Table 2.4 A Table Comparing Several IC Package Types

2.3.3: IC Substrates: Material Properties

The word "substrate" means literally, the layer ("strata") that goes underneath ("sub"). In this case, it refers to the material from which the IC is made. There are actually over twenty different materials used for making different types of ICs. These can be broken roughly into three categories:

1. Elemental

2. IV-IV compound
3. Other compounds

Table 2.5 lists some of these, along with their relevant properties.

Type	Semi-conductor	Melting Point (°C)	Density, g/cm3	Resistivity (Ω·cm)	Mobility, cm2/V·s		Dielectric Constant
					Electrons	Holes	
Elemental	Si	1417	2.33	640	1500	600	11.8
	Ge	937	5.33	10	3900	1900	16
	C (diamond)	3577	3.51	1012	1800	1600	5.5
IV-IV	SiC	>2000	3.21		325	50	10
	SiGe	Not actually used as a substrate			>2000	>800	
Other	GaAs	1237	5.31	4000	8500	400	10.9
	GaSb	712	5.62		4000	1400	15
	GaP	1465			110	75	10
	InAs	942	5.7		33000	460	14.5
	InP	1062	4.79		4600	150	14
	CdS		4.82		300	50	10
	CdTe	1092	6.06		1050	100	10.2
	PbTe	904			6000	4000	30

Table 2.5 Properties of Common Semiconductor Substrates.

Modern electronics is built on the foundation of the transistor. Transistors are made from semiconductor materials, which are materials whose resistivity ranges from approximately 10^1 to 10^8 Ω-cm. Materials whose resistivity is lower than this range are considered conductors; materials whose resistivity is higher than this range are considered insulators.

There are several properties of IC substrates that are important in making ICs from them. The following discussion of some of these properties is by no means exhaustive, but will serve as an introduction to the substrates currently in use. This introduction will also enable the reader to understand future developments in this area. There are a few terms used in this discussion that may be new to the reader. The first of these is wafer, which is a thin slice of the semiconductor material on which the IC devices (transistors, resistors, etc.) are fabricated (see Figure 2.8). The next term is monocrystalline, which means like a single (mono) crystal. Single-element solids usually take one of three forms: monocrystalline, as already defined; polycrystalline, meaning having many (poly) crystal sizes and orientations; and amorphous, meaning without (a) crystalline shapes (morph) (see Figure 2.9)

Figure 2.8 A typical 8" (200mm) silicon wafer.

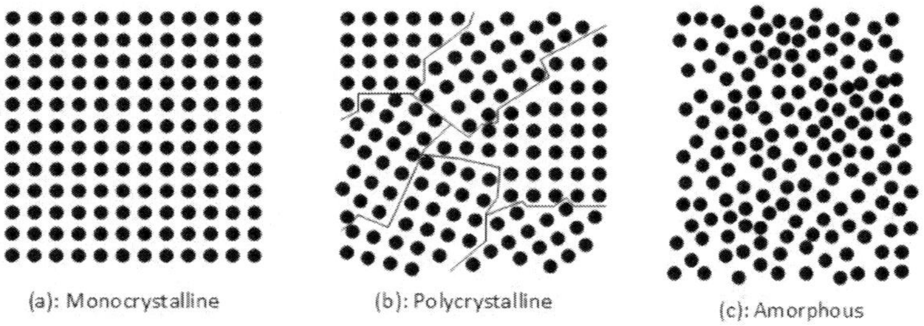

Figure 2.9 The three main forms of atomic structure in solids. (a) monocrystalline, (b) polycrystalline, and (c) amorphous crystal.

2.3.3.1: Inexpensive Material

Because transistors are made by the trillions (not at all an exaggeration), it is desirable that the raw material from which they are made be inexpensive. This means that the material should be both abundant and easy to isolate from other materials (purify).

2.3.3.2: Great Crystals

Most ICs are made from monocrystalline semiconductor material. It is therefore desirable that the material be one from which such crystals can readily be grown. "Great," in this sense, means that the crystal has very few crystal lattice irregularities and very few impurities.

2.3.3.3: Great Native Oxide

The modern process of fabricating ICs depends heavily on the ability to grow a layer of native oxide on the surface of the substrate. This means, for instance, that on a silicon substrate, we would want to grow a layer of silicon oxide, or SiO2. This layer should be easy to grow, and uniform in thickness and density.

2.3.3.4: Good Mechanical Properties

The actual electronic circuits fabricated on IC substrates in reality occupy only a small fraction of the thickness of the wafer substrate (about 0.1 percent). The majority of the thickness of the wafer substrate is necessary to prevent breakage during the manufacturing process. If the semiconductor material were less brittle, the wafer could be made still thinner, which would allow more ICs to be built with a given amount of semiconductor material. Therefore, one of the desirable mechanical properties is that it not be brittle.

Another desirable mechanical property is that it not expand significantly with temperature. Because electronic circuits create heat, each time the IC is powered up it heats up; when it is powered down, it cools down. If the semiconductor material expands significantly with temperature, it will be expanding and contracting every time the circuit is powered up and powered down. The less expansion there is, the better it is for the IC and for the IC package.

2.3.3.5: High Electron and Hole Mobility

Electron and hole mobility refers to the ease with which these subatomic particles move about within the crystalline material. The frequency at which a transistor can operate is closely related to the electron and hole mobility of the semiconductor substrate material. It is therefore desirable that the semiconductor material have a high electron and hole mobility.

2.3.3.6: Dielectric Constant

To understand the significance of the dielectric constant of the substrate material, it is necessary that we first refresh our memory of what dielectric constant refers to, which in turn requires that we refresh our memory on permittivity and capacitance.

Permittivity is defined as the degree to which a material allows the establishment of electrostatic fields. Materials with high permittivity love electrostatic fields, and readily allow them to be established within themselves. The lowest permittivity presently known is that of a vacuum, or free space, which is the absence of matter; this is defined as absolute permittivity, represented by the symbol ε_o

(epsilon naught), whose value is a constant at 8.85x10-12 F/m^2.

Capacitance is the ability to store electric charge in an electrostatic field; it is defined as:

$$C = \frac{\varepsilon_o \varepsilon_r A}{d}$$

where C= capacitance in Farads, A is the overlapping area of the capacitor plates in cm^2, d is the distance between the plates in cm, and ε_r (epsilon relative) is the dielectric constant. The dielectric constant is defined as the ratio between the permittivity of the material and absolute permittivity, and is thus a unitless number. Another, perhaps easier way to remember what the dielectric constant is, would be to think of it as how much better the material is, compared to free space, at allowing the establishment of electrostatic fields within that material. Thus a material with a dielectric constant of 5 would be five times better at allowing the establishment of electrostatic fields within it. The topic of capacitance and capacitors will be covered in more detail in chapter 5.

Now that we remember these matters of permittivity, capacitance, and dielectric constant, the next step is to jog our memory as to the significance of capacitance in modern integrated circuits. Electronic circuits are made of many, many stages; each stage brings signals into its inputs, processes these signals in various ways, and then passes these signals on to the next stage or stages of the circuit. It is desirable that the thousands and millions of signals in an IC not become mixed with each other as they are passed from stage to stage. However, parasitic capacitance is always present when there are two or more signals, and because the reactance (opposition to AC current) of capacitors is inversely proportional to frequency, and capacitance is directly proportional to the dielectric constant, this is where the problem arises.

For example, two signals running next to each other across a portion of the IC are separated by an insulator, usually SiO_2. This insulator acts as the dielectric between the plates of the capacitor (which plates are the conductor lines of the signals). If the insulating material has a high dielectric constant, there will be more capacitance, resulting in lower reactance, which in turn results in greater crosstalk, or signal interference, between these two signals.

In summary, the higher the dielectric constant of the substrate material, the greater the problem of crosstalk between adjacent signals. Therefore it is desirable that the substrate material have a dielectric constant as low as possible. The lowest possible dielectric constant is when $\varepsilon_r = \varepsilon_o$, which would make $\varepsilon_r = 1$. Typical values of the dielectric constant for substrate materials range from 5.5 to as high as 30, as seen in Table 2.5.

2.3.4: IC Substrates: Materials

As mentioned previously, IC substrate materials can be classified as the following:

1. elemental
2. IV-IV compound
3. other compounds

Each category has some materials which are in significant use, and others which are experimental or in very limited production. The most common elemental substrates, in order of their significance in modern electronics, are silicon (Si; element #14), germanium (Ge; element #32), carbon (C; element #6), and tin (Sn; element #50). All of these are found in column IVA of the periodic table, and thus are commonly referred to as "column 4" materials. Silicon is, by every measure, overwhelmingly predominant in making ICs, easily accounting for over 99.9% of all ICs. Germanium is significant in that the first transistor ever was made of it, and germanium continues to be used in modest amounts for certain applications.

Carbon is an interesting entry in this list; its monocrystalline form (diamond) is well known for its applications in jewelry and industrial abrasives and cutting, but it is not yet well established in the IC market. However, because of some remarkable properties which it possesses, research has shown promise, and it is very likely to be developed much further in the coming years.

The second classification (IV-IV Compound) refers to substrate materials that are made from two "column 4" elements. There are presently only two significant members of the second classification: SiGe and SiC. SiGe is significant primarily in the area of high-frequency circuits (multi-GHz). SiC is significant in making LEDs that emit blue light, and in making ICs that operate at very high temperatures (>200°C).

The third classification (Other Compounds) has many members (see Table 2.6), most of which are used only in highly specialized applications. One that has relatively widespread use in the area of high-frequency circuits (multi-GHz) and LEDs is GaAs, gallium-arsenide. A few of the other substrates may be familiar to the reader in applications such as photosensitive resistors (CdS, cadmium sulfide); LEDs (a variety of compounds including InP (indium phosphide) and InAs (indium arsenide)); and infrared sensors (CdTe, cadmium telluride).

CHAPTER SUMMARY

There are multiple levels of packaging in electronic products; each level must be considered separately. The second level of packaging includes integrated circuits, which come in three families (analog, digital, and mixed); each family contains many types of functional blocks, and there are literally thousands of choices for each type of functional block.

IC packages are level 1, and serve many purposes, some of which are conflicting. Since there is no IC package which is ideal for all applications, the correct choice is a result of carefully considering the tradeoffs in a given application. IC packages for consumer products are generally those which are the least expensive.

Silicon is, by far, the predominant substrate material used for making ICs, but it is certainly not the only choice. If silicon cannot perform to the needed level for a given application, there are many other substrate material options available, although all of them are much more expensive than silicon. There are certain characteristics which make a substrate material desirable; several IC substrate materials were compared for their performance in these characteristics, as summarized in Table 2.6.

FOCUS QUESTIONS

1. List in order the first 5 packaging levels in electronics
2. Why are there so many choices available for each type of functional block of IC?
3. Which IC package purpose(s) would be most important for electronic wristwatches?
4. Which IC package purpose(s) would be most important for a modern microprocessor?
5. Which IC package type(s) would you expect to see most of in a boom box? Why?
6. GaAs is the second-most-common IC substrate material used, mainly for very high-frequency applications (>1 GHz). What characteristic of GaAs makes it good for this application?
7. Which of the three forms of solid matter are most common in IC substrates? Why?
8. Explain why we want a low dielectric constant for a good IC substrate material.

Material	Symbol	Cost	Crystal Quality	Quality of Oxide	Mechanical Properties	Electron/Hole Mobility (cm2/Vs)	Dielectric Constant
Carbon	C	High	Medium		Superb	1800/1600	5.5
Germanium	Ge	Medium	High	Low	Good	3900/1900	16
Silicon	Si	Low	Very High	Very High	Good	1500/600	11.8
Silicon-Carbide	SiC	Medium	Medium	Medium	Good	400/50	10
Silicon-Germanium	SiGe	Medium	High	Medium	Good	Very High	
Boron-Nitride	BN	High	Low				
Boron-Phosphide	BP	High	Low				
Cadmium-Sulfide	CdS	High	Low			300/50	10
Cadmium-Selenide	CdSe	High	Low			800/	10
Cadmium-Telluride	CdTe	High	Low		Poor	1050/100	10.2
Gallium-Antimonide	GaSb	High	Low			4000/1400	15
Gallium-Arsenide	GaAs	High	Low	Low	Poor	8500/400	10.9
Gallium-Phosphide	GaP	High	Low			110/75	10
Indium-Antimonide	InSb	High	Low			78000/750	17
Indium-Arsenide	InAs	High	Low			33000/460	14.5
Indium-Phosphide	InP	High	Low			4600/150	14
Lead-Sulfide	PbS	High	Low			600/700	17
Lead-Telluride	PbTe	High	Low			6000/4000	30
Zinc-Sulfide	ZnS	High	Low			165/	8

Table 2.6 Table Comparing Various Properties of IC Active-Layer Materials

FOR ADDITIONAL READING

Blackwell, Glenn R., "The Electronic Packaging Handbook", Chapter 3, CRC Press, Boca Raton, FL, 1999.
Harper, Charles A., "Electronic Packaging and Interconnection Handbook", Chapter 6, McGraw-Hill, New York, NY, 1997.
Harper, Charles A., and R.M. Sampson, "Electronic Material and Processes Handbook", McGraw-Hill, New York, NY 1994.
Lau, John H., C.P. Wong, J.L. Prince, and W. Nakayama, "Electronic Packaging: Design, Materials, Process, and Reliability", McGray-Hill, New York, NY 1998.
Matisoff, Bernard S., "Handbook of Electronics Packaging Design and Engineering", Chapter 4, Van Nostrand Reinhold, New York, NY, 1990.
Pecht, Michael, "Handbook of Electronic Package Design", Chapter 2, Marcel Dekker, New York, NY, 1991.
Pecht, Michael, R. Agarwal, P. McCluskey, T. Dishongh, S. Javadpour and R. Mahajan, "Electronic Packaging: Materials and Their Properties", CRC Press, Boca Raton, FL, 1998.
Seraphim, Donald P., R.C. Lasky, and C. Li, "Principles of Electronic Packaging", McGraw-Hill, New York, NY, 1989.
Tummala, Rao R. and Eugene J. Rymaszewski, "Microelectronics Packaging Handbook", Van Nostrand Reinhold, New York, NY, 1989.

CHAPTER 3: FIXED AND VARIABLE RESISTORS

CHAPTER OVERVIEW

This chapter covers the following concepts:

- The ideal resistor has several desirable characteristics
- There are many types of resistors, each with its own attributes; no resistor type is ideal in every way
- Seven of the most common resistor types are described, along with their characteristics and how they are made
- Potentiometers and rheostats are the types of variable resistors, but differ significantly in their application
- The three main types of variable resistors are described

3.1: FIXED RESISTORS

Many times in the design of a circuit, the designer puts in resistors as needed, and doesn't think much more about them. After all, an Ohm is an Ohm, right? However, the choice of the right resistor type is often a key to a successful circuit. The purpose of this section is to enable manufacturing engineers to understand the various types of resistors that exist in industry.

The main types of fixed resistors are IC, carbon composition, carbon and metal film, wirewound, network and chip and MELF (see section 3.1.9 below). The characteristics of each of these resistor types will be discussed in this section. But first, we should become familiar with the operation and function of fixed resistors, and the characteristics we would want in a good resistor.

3.1.1: Resistor Operation and Functions

All materials, except for superconducting materials, offer some opposition to the flow of electricity. Electrons, which make up the current in a circuit, do not march down the conducting path in lock-step, but rather drift along from one atom to another while orbiting atoms at rates of about 10^{15} Hz. As they drift along under the influence of an electromotive force (EMF, or voltage), they encounter opposition in the form of other electrons, atoms, or molecules which take some of the electrons' energy and convert it to localized motion, which we know as heat.

The purpose of resistors is to provide a known and fixed amount of opposition (resistance) to electrical current flow. The main reasons such resistance is needed is to:

1. limit or divide the current, or
2. drop a known amount of voltage.

In these functions, resistors serve a very necessary purpose in electronic circuits, and are the second most common electronic circuit element, behind the transistor or IC.

3.1.2: Resistor Characteristics

Let's consider the characteristics of a good resistor in alphabetical order, and let's take for our example a resistor which is rated at 1000 Ω (1 kΩ). A good 1000-Ω resistor will measure 1000 Ω today, tomorrow, next year, and next decade. That is, it will not exhibit *age drift*. A good 1000-Ω resistor will not add a great deal of expense to our circuit; that is, it will have a *low cost*.

Occasionally, electronic circuits subject resistors to current surges. As the current surge passes through the resistor, it creates a power dissipation surge. Therefore, we would want our resistor to be able to handle these *current surges*.

If our resistor is as good as we need it to be, it will not add anything to our circuit other than the 1000 Ω we wish it to add. That is, it will have very *low parasitics* (specifically, low parasitic capacitance and inductance).

Our 1000-Ω resistor would, ideally, last forever; at least, we would like it to last for a very long time. In other words, it should have very *high reliability*. And because we always seem to like our electronic circuits small, our resistor should be *small in size*.

We would also like our 1000-Ω resistor to measure 1000 Ω at room temperature, at -30°C, at 75°C, and at essentially all temperatures for which we want to use the circuit. That is to say, we want the resistor to have a very *low temperature drift*. We would also like our 1000-Ω resistor to measure exactly 1000.000...Ω. That is, it would have a *very low tolerance*. This is a difficult wish to fulfill in reality, since resistors that are exactly 1000.000...Ω are very hard to make. However, the closer the resistors are to their rated value, the lower their tolerance is, and the closer they are to what we originally wanted.

By now, the reader has probably realized that it will not be easy to find a resistor that fulfills all of the above desired characteristics. However, the various types of resistors that exist each have their strengths, as will be discussed below.

3.1.3: Integrated Circuit Resistors

Although they take up more area than transistors, integrated circuit resistors are very easy to make, and are therefore the least expensive of all the resistor types. This is primarily because millions or billions can be made at the same time. The main types of integrated circuit resistors are diffused, implanted, and thin film. All of these types are made by processes which will be covered in chapters 7, 8 and 9.

Diffused resistors typically have a very poor tolerance (±10 - 20%) and a relatively limited range of values (100 Ω - 100 k Ω), but are the easiest to fabricate. Implanted resistors have a better tolerance (±2 - 10%), and a better range of values (50Ω - 500kΩ). Thin film resistors have excellent tolerance (±1 - 5%; trimmable to .01%) and a wider range of values (10 Ω - 1 M Ω), but are more complex to fabricate, especially if trimming is required.

All types of IC resistors exhibit excellent *tracking*, because they are fabricated in parallel with all the other resistors on the IC chip. That is, if a given resistor is supposed to be 1000 Ω, but when fabricated (due to tolerances) comes in at 1050 Ω (+5%), all the other resistors on the IC chip will be extremely close to the same +5% values (they "track" each other). By taking advantage of this fact, most circuit designs can take advantage of the close tracking of IC resistors and suffice with their relatively poor tolerances.

One significant limitation with IC resistors is the relatively narrow range of values that can easily be fabricated on the IC. If a circuit requires a range of greater than about 100:1, the cost of the IC goes up. Thus, if it is necessary to have both 100-Ω resistors and 1-M Ω resistors on the same IC (a range of 10,000), it will result in additional cost.

Another significant limitation for IC resistors is the very low power they can dissipate. Power dissipation capability is directly proportional to surface area, and thus to size; IC resistors are, necessarily, very small. Thus, if power ≥1/8 W will be dissipated, an IC resistor will generally not suffice.

3.1.4: Carbon Composition Resistors

Carbon composition resistors, or "carbon-comp" as they are often called, are the oldest resistor type, and up until the 1980s they were the predominant type as well. As shown in Figure 3.1, they are composed of a small cylinder of resistive material, made from a mixture of powdered carbon (graphite), a powdered insulating material (often a powdered ceramic such as alumina), and a binder to keep the molded material in the desired shape. This cylinder of resistive material is formed in a mold that also has the leads in it; as the three elements of the resistive material (graphite, alumina, and binder) are introduced into the mold and the mold is pressed and heated, the leads become permanently bonded to the resistive cylinder. The leads and resistive cylinder assembly are then covered with a thick layer of epoxy paint for protection; this epoxy also forms the final shape of the resistor. The final step is to paint the value and tolerance stripes on the resistor.

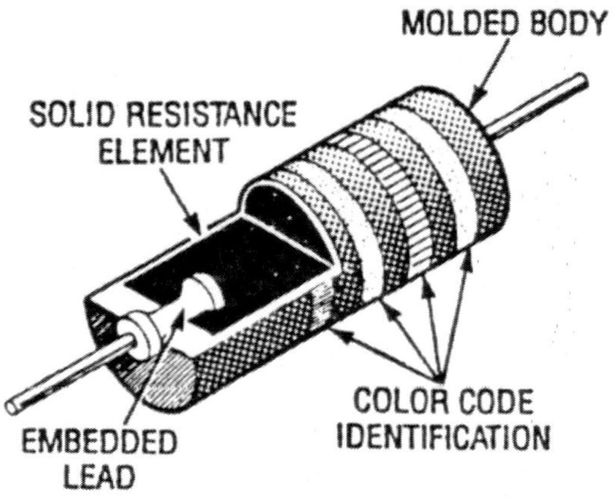

Figure 3.1 Carbon composition resistor construction. *Courtesy of R. M. Marston.*

The size of the resistive cylinder is determined by the power rating of the resistor, and not by the resistance value. In order to achieve a wide range of resistor values, the mixture ratio of the graphite and the powdered ceramic is varied; resistors with a low resistance value have a high graphite/ceramic ratio, while those with a high resistance value have a low graphite/ceramic ratio. Values from 10 Ω to 30 MΩ are standard, with other values available. Different wattage ratings are available, with ⅛ W, ¼ W, ½ W, 1 W and 2 W being standard.

The carbon comp resistor is simple to manufacture and is made from inexpensive materials in highly automated facilities. Thus, one of its greatest strengths traditionally has been its low cost. It is also a very rugged resistor, with both high reliability and excellent handling of current surges. Carbon comp resistors also have relatively low parasitics. The weaknesses of the carbon comp resistor include a relatively poor tolerance (±5% to ±20%), and relatively poor age and temperature drift. For these reasons, in addition to the fact that the cost of carbon and metal film resistors has dropped, the carbon comp resistor has declined significantly in usage, with most designs turning instead to the carbon and metal film, network, or chip resistors.

3.1.5: Carbon and Metal Film Resistors

From the outside, it is essentially impossible to distinguish a carbon film resistor from a metal film resistor. Both are made by the same process, which begins with a ceramic mandrel usually made from Al_2O_3 (alumina) (see Figure 3.2). This mandrel is then coated with an ink ("film"), in the case of thick-films, or a vacuum-deposited layer in the case of thin-films. Thick-film inks are a mixture of either carbon powder or metal powder (usually ruthenium oxide), along with a powdered insulating material (again, usually alumina), and fluids to make an ink of it. Thick-film layers are typically thicker than 10 μm. This film is then cured in an oven, after which the resistor is laser-cut to its target value. Thin films are usually made of tungsten; the layer is typically thinner than 1 μm; thin films are also laser-cut.

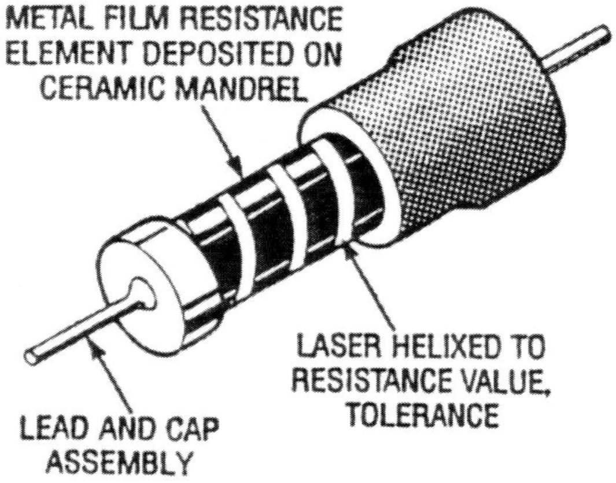

Figure 3.2 Metal film resistor construction. *Courtesy of R. M. Marston.*

The laser ablates (burns away) the film in a helical pattern; some of the mandrel is also ablated in the process, as can be seen in Figures 3.3 and 3.4. Leads are then attached to the mandrel by a soldering process, after which the assembly is dipped into epoxy paint. The final step is to paint the resistor value and tolerance stripes on the resistor.

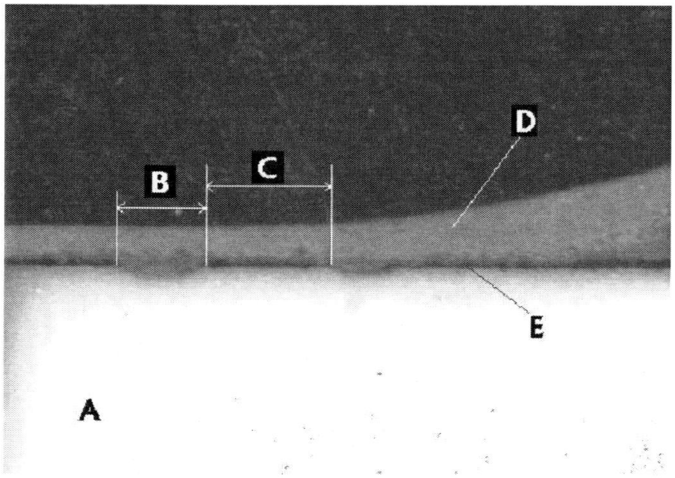

Figure 3.3 Cross section of a metal or carbon film resistor, showing (a) the ceramic mandrel, (b) the width of the ablation, (c) the width of the resistance element, (d) the epoxy overcoat, and (e) the resistance film deposited on the mandrel.

Figure 3.4 A metal carbon film resistor with part of the conformed coat removed.

The size of the mandrel is the primary factor which determines the power rating of the resistor. A relatively wide range of resistance values can be laser-cut onto the same film-covered mandrel, but to make the multiple decades range needed by designers, multiple ink mixtures are used. Resistors with high resistance values are made from an ink with a low ratio of carbon or metal powder to ceramic powder; conversely, resistors with low resistance values are made from an ink with a high ratio of carbon or metal power to ceramic powder. Values from 10 Ω to 33 M Ω are standard, with wattage ratings from ⅛ W to 1 W being standard.

Carbon and metal film resistors are relatively simple to fabricate using highly automated facilities. Because there are more steps and more materials involved, they are typically more expensive than carbon comp resistors. Recently, however, this difference has diminished until the difference is, in many designs, insignificant.

Because they are laser-trimmed to their design values, carbon and metal film resistors have an excellent tolerance. Tolerances of ±1 to 5% are standard, with precision values available (for higher cost) to tolerances of ±.01%. Also, because of the type of construction and the materials used, carbon and metal film resistors exhibit lower age and temperature drift than carbon comp resistors.

The helical pattern does create a small problem; it resembles a small inductor, which indeed it is.

Because each turn of the helical pattern is separated from the previous one by the insulating substrate, there is also additional capacitance. Therefore, carbon and metal film resistors have more parasitic inductance and capacitance than carbon comp resistors.

There is another problem created by the manner in which these resistors are constructed, and that is their ability to handle current surges. The cross-sectional area of the resistance material in which the current flows (the thin helical-trimmed film on top of the mandrel) is much smaller than in the carbon comp resistors. Thus, when a current surge passes through a carbon or metal film resistor, it can create much higher instantaneous heat gradients within the resistor; these heat gradients can quickly damage the resistor. The result is that carbon and metal film resistors are poorer than carbon comp resistors at handling current surges. This, along with the fact that there are more parts in a carbon or metal film resistor, makes their reliability somewhat lower than that of carbon comp resistors.

3.1.6: Network Resistors

Network resistors are made from the same materials as carbon or metal film resistors, but in a different form factor. Instead of depositing the thin or thick film on a mandrel, it is deposited on a rectangular slab of alumina (see Figure 3.5); multiple resistive elements are formed on the same ceramic substrate. The leads are attached as on the carbon and metal film resistors. The resistive elements are laser trimmed with a small incision, as shown in Figure 3.6. The final step is to coat the assembly with epoxy paint.

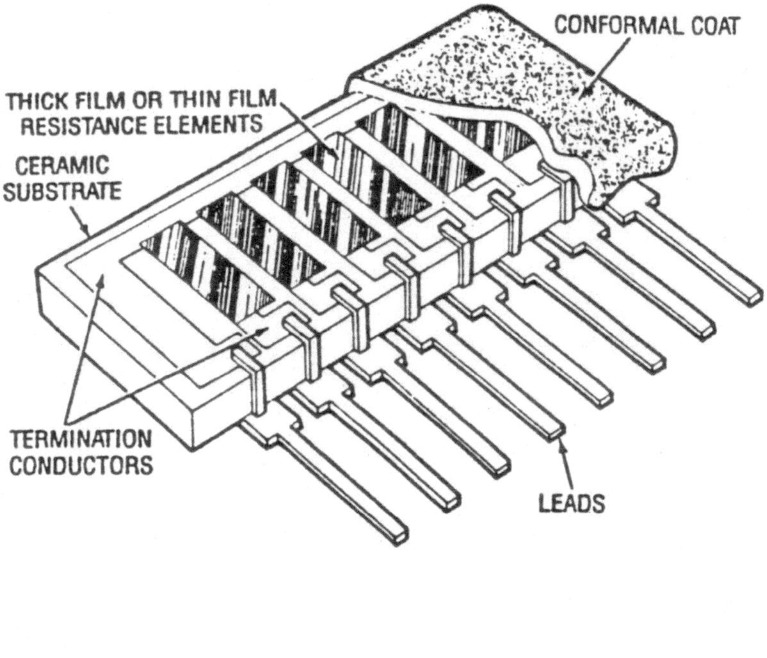

Figure 3.5 Network resistor construction. *Courtesy of R. M. Marston.*

Figure 3.6 Top view of a network resistor with the epoxy coating removed.

This form of resistor is convenient for large numbers of resistors, because they are smaller and often less expensive than their discrete counterparts. They also usually have a lower overall failure rate (higher reliability) than the combined failure rate of the multiple discrete resistors they replace. Also, because they are made at the same time on the same substrate, they exhibit excellent tracking, both in initial value and in temperature drift.

Because network resistors are laser-trimmed, they exhibit a tolerance range almost identical to that of carbon and metal film resistors, or ±1 to 5% standard with precision values available (for higher cost) to tolerances of ±.01%. Also, because the laser trimming pattern is non-helical, they exhibit lower parasitic inductance and capacitance than carbon and metal film resistors.

As seen in Figure 3.7, network resistors are available in both through-hole (SIP) form and surface-mount form (SOIC and LCC). Many standard values and configurations are available; custom network resistors are also very common, especially in high-volume applications. One challenge of this form of resistor is the limited range of values available on the same substrate (without additional cost). Ideally, all the resistive elements on the substrate are made from the same ink, and are then laser-trimmed to value. However, this limits the range of values that the resistive elements can have. Typical is a limitation of a 100:1 ratio. Wider ratios require more than one ink, which increases the number of process steps and thus the cost.

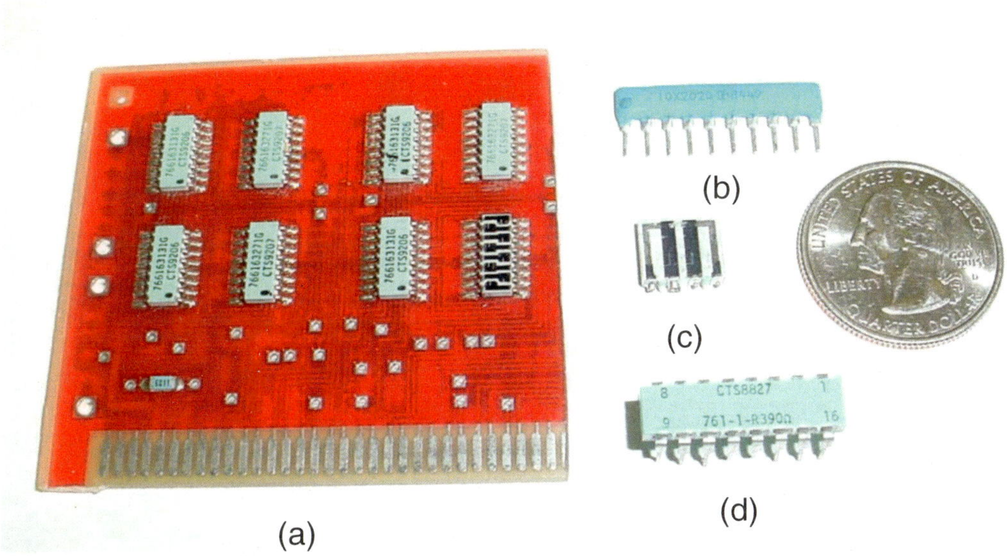

Figure 3.7 Examples of network resistors. (a) SMT type on PWB, one has epoxy coating removed, showing the resistance ink elements, (b) SIP type, (c) SIP type with epoxy coating removed, showing resistance ink elements, d) DIP type.

3.1.7: Wirewound Resistors

Wirewound resistors are made from a coil of resistive wire, usually nickel-chromium ("nichrome"), as seen in Figure 3.8. This is the same kind of wire which is used in electric heater elements, the part of the heater that glows red when electricity passes through it. The nichrome wire is wrapped around a mandrel usually made of alumina. The end-caps are attached by crimping or soldering, and the assembly is then coated with epoxy paint.

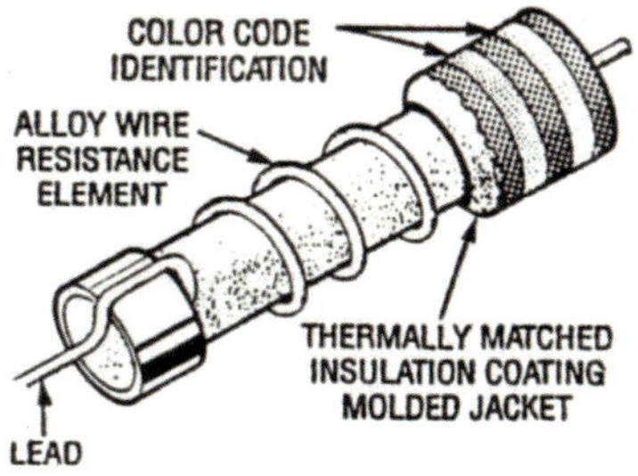

Figure 3.8 Wirewound resistor construction. *Courtesy of R. M. Marston.*

Wirewound resistors come in two main types: power and precision. The purpose of power wirewound resistors is to provide high values of power dissipation. Standard values range from 5W to over 100 W. Precision resistors are, as may be supposed, for high-precision applications, where the actual value of the resistor must be very close to the rated value. Standard tolerances range from ±1% down to ±.01%; specialty companies offer them down to ±.001%. They are excellent for low age drift. Precision wire-wounds are also excellent for low temperature drift, but power wire-wounds are not.

Wirewound resistors are the most expensive type; they are only justified where the tolerance or power ratings of the other types are inadequate. They also exhibit significant parasitic inductance and capacitance, as can be supposed by their coiled construction. Where low parasitic inductance is essential in a wire-wound resistor, bifilar-wound resistors are available. As shown in Figure 3.9, these wirewound resistors use a doubled-over wire element; the currents that flow in the wire flow in opposite directions, canceling out most of the inductive effects.

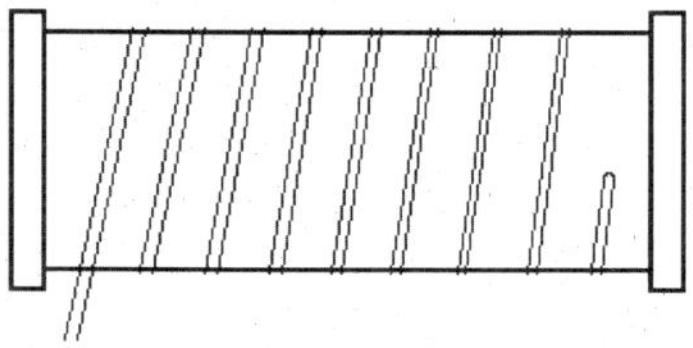

Figure 3.9 Bilfilar windings on a mandrel.

Wirewound resistors also have the poorest reliability of all resistor types, especially power wirewounds. This is partly due to the heat they put out when dissipating their rated power.

3.1.8: Chip Resistors

Chip resistors are made by the same basic process by which network resistors are made, and from the same materials. As seen in Figure 3.10, a small slab of alumina is coated on one side with the thin or thick film, the end caps are attached, and the resistive element is laser trimmed. Being a surface mount technology, chip resistors have become predominant in modern electronic circuits. They are the least expensive of all resistor types, by far the smallest, and exhibit excellent tolerances (±5% down to ±.5% being common). They are also the lowest in parasitic inductance and capacitance, and have an excellent reliability rating.

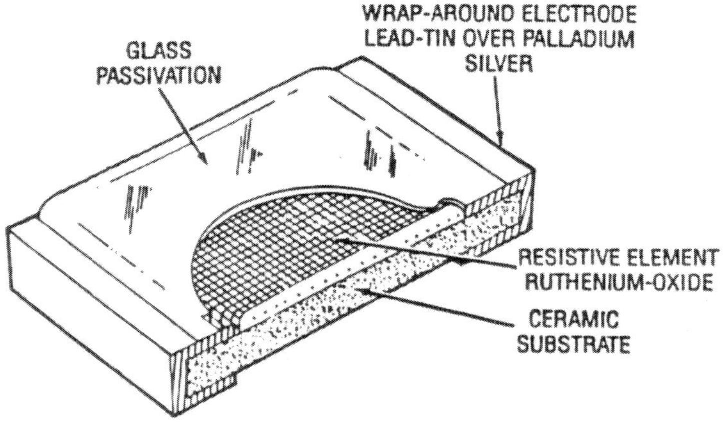

Figure 3.10 Chip resistor construction. *Courtesy of R. M. Marston.*

Chip resistors are very small, which means they have a very low power rating, but this has not slowed their popularity. Initially, chip resistors came in sizes of 180 mils by 120 mils (the 1812 style) and 120 mils by 60 mils (1206 style), but this has been extended downward, due to continued pressure to make circuits smaller. Today they also come in styles 0805, 0603, 0302, 0201, and even 01005. The 01005 style is so small (only 10 mils by 5 mils) that to drop one is to lose it forever. Keep in mind that the human hair is only about 2 mils thick; a typical grain of salt is about the same size as an 01005.

3.1.9: MELF Resistors

MELF (Metal ELectrode Face) resistors were introduced about the same time as chip resistors, and have remained a viable option, even though of late their popularity has been declining. The process for making a MELF is very similar to that for making a film resistor (section 3.1.5), except that the terminals on the ends are lacking leads (see Figure 3.11). The cylindrical mandrel is still the same alumina material; the resistive layer is made from the same materials. The characteristics are in nearly every way identical; the main difference is simply that the MELF is a surface-mount device, while axial film resistors are not.

3.1.10: Resistor Summary

Discrete resistors are used where integrated circuit resistors cannot satisfy the need. This usually is because a wide range of values is needed, or higher precision, or higher power. Fortunately, there is a wide range of discrete resistor types available to meet this need, each type with its own strengths and weaknesses. Table 3.1 summarizes these characteristics, and represents typical values. Specialized resistors exist for many requirements outside the typical values.

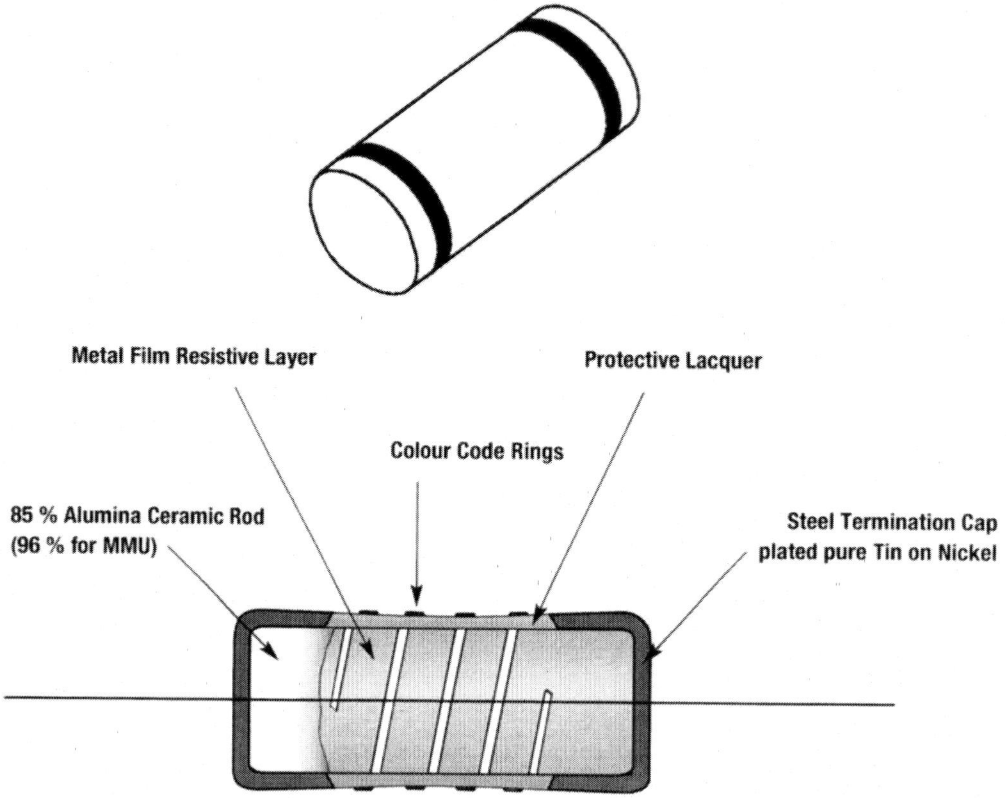

Figure 3.11 Metal Electrode Face (MELF) resistor construction. *From www.st.com/stonline/books and www.beyschlag. Delbey-eng/pro. Courtesy of BCComponents. BV.*

3.2: VARIABLE RESISTORS

Sometimes in the operation of electronic circuits, it is desirable that a certain current or voltage be variable. One of the simplest ways to accomplish this is through a variable resistor. The two main types of variable resistors are ceramic substrate and wire-wound.

Resistor Type	Resistance Range (Ohms)	Tolerance Range (%)	Power Range (Watts)	Temperature Drift (ppm/°C)	Age Drift (%/kHrs)	Failure Rate (Reliability) (%/kHrs)	Cost
Integrated: Ion Implanted	5 - 1M	±1 - ±20	<.1	±100 - ±200	±0.01 - ±20	.00001	lowest
Carbon Comp	1 - 100 M	±5 - ±20	0.1 - 5	+200 - +2000	±5 - ±10	.0002	low
Carbon/ Metal Film	1 - 100 M	±0.1 - ±5	0.1 - 5	+50 - +200	±0.5 - ±3 / ±0.1 - ±0.5	.0033	low
Network	5 - 30 M	±0.1 - ±5	0.1 - 1	+50 - +200	±0.1 - ±0.5	.002	medium
Wire-wound: Power Precision	0.01 - 10 k 0.01 - 100 k	±2 - ±10 ±0.001 - ±1	5 - 1500 0.25 - 2	+200 - +500 +0.5 - +50	±2 - ±5 ±0.1 - ±0.5	.01 .001	high high
Chip	5 - 10 M	±0.1 - ±5	0.01 - 0.1	±50 - ±200	±0.1 - ±0.5	.0004	very low
MELF	1 - 50 M	±0.1 - ±5	0.1 - 1.0	±50 - ±200	±0.1 - ±0.5	.003	low

Table 3.1 A comparison of the characteristics of common resistor types.

3.2.1: Operation and Functions

Variable resistors perform two main functions: they provide a variable current, or a variable voltage, through the operation of a variable resistance. Variable resistors that provide a variable current are termed rheostats, from the Greek, rheos, meaning flowing, plus stat, short for state or condition. The correct symbol for a rheostat is shown in Figure 3.12(a). Note that they need only have two terminals.

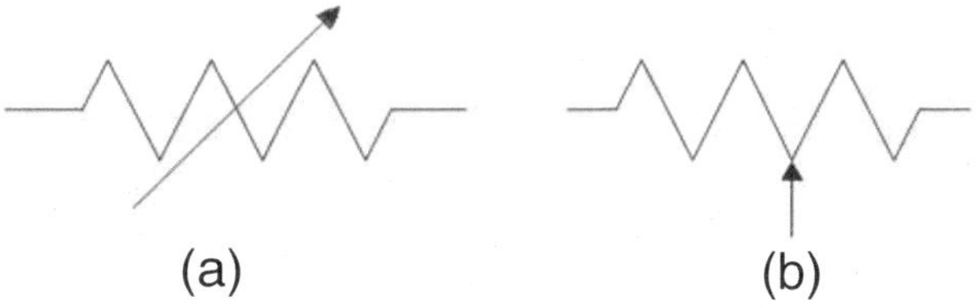

Figure 3.12 Symbols for variable resistors. (a) the correct symbol for a rheostat, (b) the correct symbol for a potentiometer.

Variable resistors which provide a variable voltage (or "potential") are termed potentiometers, or usually just "pots". The correct symbol for a potentiometer is shown in Figure 3.12(b). Note that potentiometers must have three terminals. Most electronic circuits which use variable resistors use them as potentiometers; for this reason, the term "pot" is commonly used to designate variable resistors in general, regardless of their function in the actual circuit. However, there is usually a significant difference between potentiometers and rheostats.

Producing a variable voltage usually does not require much in the way of power dissipation. For this reason, most potentiometers are small, and some are truly tiny (see Figure 3.13). However, this is usually not true for rheostats.

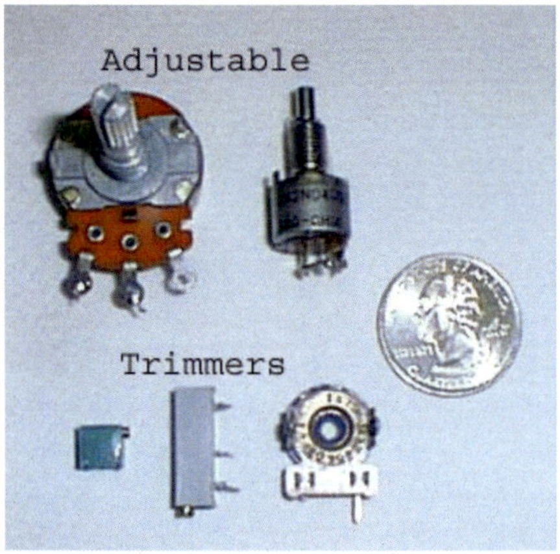

Figure 3.13 Examples of adjustable and trimmer potentiometers.

The power dissipated in a given resistance is given by the equation $P=I^2R$; note the squared term after the I (current). If a variable resistance is used to provide a variable current, this current often causes significant power dissipation in the rheostat. For this reason, most rheostats are significantly larger than potentiometers, and some only have two terminals, since that is all they must have.

A word of caution is in order in the application of rheostats. Variable resistors usually go from a low of 0 Ω to a high of their rated value. If a 1000-Ω rheostat is used to control a current from a 10-V source, the current may vary from a low of (10 V / 1000 Ω = 10 mA) to a high of (10 V / 0 Ω = ∞ A). Clearly an infinite current is not possible, but what is possible is that when the rheostat is adjusted to its lowest value, a very large current flows. This large current is the problem, and many electronics engineers learn by sad experience to put a separate limiting resistor in series with the rheostat to prevent such a fault current (see Figure 3.14).

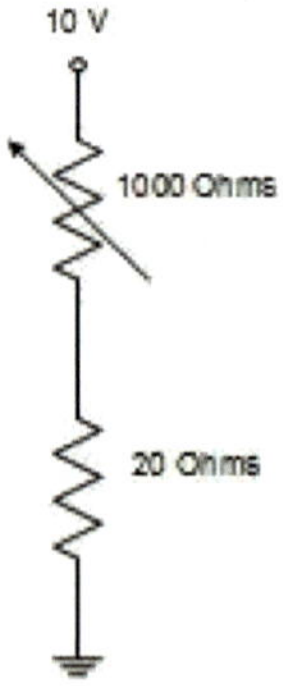

Figure 3.14 Use of a limiting resistor (20 Ohms) in series with a rheostat to limit the maximum current through the rheostat.

3.2.2: Ceramic Substrate

At present, potentiometers are almost entirely of this type. The basic parts of a ceramic substrate potentiometer are shown in Figure 3.15. The ceramic substrate is inked with a strip of resistive ink, as in fixed network and chip resistors. The strip of resistive ink may be in a circular pattern or in a straight line. The ink may be made of ceramic and carbon for a carbon film potentiometer, or of ceramic and metal, for a "cermet" potentiometer. The carbon film or cermet strip is contacted by a wiper, which makes contact against it as it moves along the strip. The movement of the wiper is controlled by the shaft of an adjustable potentiometer, or by an adjusting screw or slot in a trimmer potentiometer.

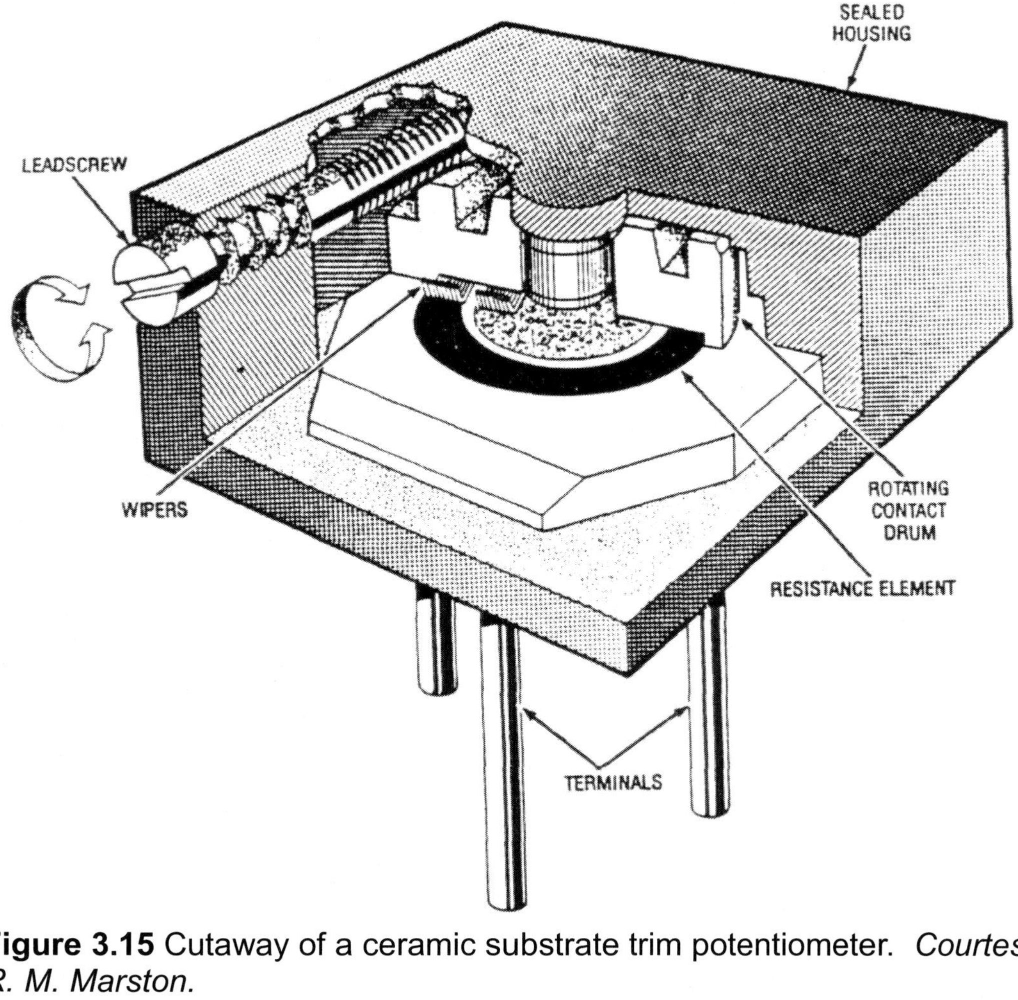

Figure 3.15 Cutaway of a ceramic substrate trim potentiometer. *Courtesy of R. M. Marston.*

A trimmer potentiometer, or "trim pot," is set in the factory and is then usually sealed. It is not intended to be adjusted by the end user of the product. An adjustable potentiometer is the opposite; its shaft protrudes through an opening in the enclosure, and a knob is attached, so that the potentiometer may be adjusted by the end user. The volume dial or knob on an inexpensive clock radio is an example of this type of variable resistor.

Trim pots are only intended to be adjusted a few times, and are not designed for long life. In those few applications where a trim pot will be adjusted many times, a special long-life type should be found and used. Adjustable potentiometers, on the other hand, are designed for long life, and can easily endure hundreds of adjustment cycles. However, even these do not last as long as many products need them to, as evidenced by how many of us are familiar with the symptoms of worn potentiometers. These symptoms include dead spots and noisy spots as the volume potentiometer is adjusted, a characteristic scratchy sound. The main causes of this wear are dust and the spring force necessary to keep the wiper in contact with the resistive ink strip. Dust and the spring force act like sandpaper on the resistive ink, significantly shortening its life.

3.2.3: Wirewound

Wirewound potentiometers and rheostats are much more expensive (two to ten times more) than the ceramic substrate variety, but the cost buys a device which lasts much longer and is capable of handling much more current. An example of a wire-wound potentiometer is shown in Figure 3.16. As can be seen, a coil of resistive wire (often nichrome) is contacted by the wiper; as the potentiometer is adjusted, the wiper moves along the side of the wire coil, providing the variable resistance. There is no thin ink layer to wear out, only the coil of wire and the wiper. The wire is much thicker than the ink, and is thus capable of carrying much more current. Almost all rheostats are of this variety.

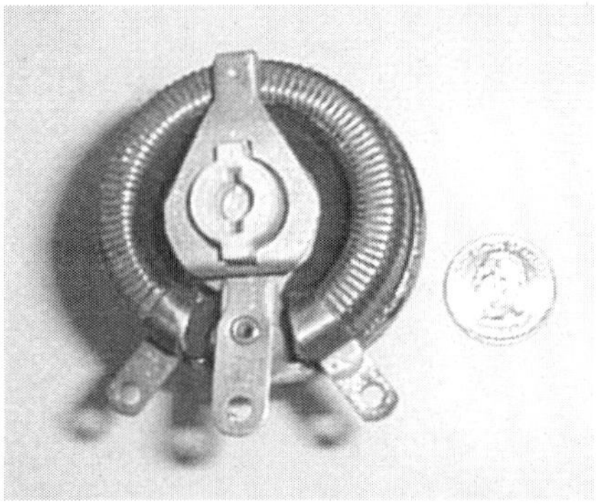

Figure 3.16 An example of a three-terminal wirewound potentiometer (rheostat).

CHAPTER SUMMARY

Contrary to common belief, there is a significant difference in the performance of the several types of resistors. There are several desirable characteristics for a resistor, including low age drift, low cost, ability to handle current surges, low parasitics, high reliability, low temperature drift, and a low tolerance. There are seven common types of fixed resistors, and each type has its respective weaknesses and strengths in each of these areas. These characteristics were compared in Table 3.1.

Variable resistors come in two basic types: rheostats for varying current, and potentiometers for varying voltage. They are different in their application and their construction, but share some common characteristics. Another way of categorizing variable resistors is by whether they are intended to be adjusted only a very few times in the factory (trimmers), or whether they are intended to be adjusted many times by the user (adjustable). If using a potentiometer as a rheostat, it is wise to use a limiting resistor to prevent the potentiometer from being accidentally overheated.

FOCUS QUESTIONS

1. What are the two main functions of resistors?
2. What are seven of the desirable characteristics of a fixed resistor?
3. Describe what temperature drift is for a resistor.
4. How much capacitance does an ideal 10.00 kΩ resistor have?
5. Integrated circuit resistors are far from ideal in most characteristics, yet they are by far the most common type of resistor made today. Why is this so?
6. What are the two main strengths of the carbon-composition resistor?
7. How are film resistors better than carbon-composition resistors?
8. What are two of the main advantages of network resistors? What is one of the weaknesses of this type of resistor?
9. What are two of the main advantages of wire-wound resistors?
10. Which type of discrete resistor is the smallest?
11. Describe the main differences between a potentiometer and a rheostat.
12. What is the basic difference between a trim pot and an adjustable pot?

FOR ADDITIONAL READING

Elliott, David J., “Integrated Circuit Fabrication Technology”, Chapter 1, McGraw-Hill, New York, NY 1989.

Harper, Charles A., “Electronic Packaging and Interconnection Handbook”, Chapter 6, McGraw-Hill, New York, NY, 1997.

Harper, Charles A., and R.M. Sampson, “Electronic Material and Processes Handbook”, McGraw-Hill, New York, NY 1994.

Marston, Ray, “Resistors”, Electronics Now, Feb 1993, p. 57-63,90.
Pecht, Michael, “Handbook of Electronic Package Design”, Chapter 2, Marcel Dekker, New York, NY, 1991.
Sinclair, Ian Robertson, “Passive Components: A User’s Guide”, Oxford, 1990.
Wolf, Stanley, and R.F.M. Smith, “Student Reference Manual for Electronic Instrumentation Laboratories”, Chapter 10, Prentice-Hall, Englewood Cliffs, NJ, 1990

CHAPTER 4: FIXED AND VARIABLE CAPACITORS

CHAPTER OVERVIEW

This chapter covers the following concepts:

- There are five main families of capacitors, based on the type of dielectric used
- How a capacitor functions, and the formula for a parallel-plate capacitor
- The desirable characteristics of a capacitor
- Dielectric strength, dielectric constant, and dissipation factor
- A comparison of the most common types of capacitors
- The basic construction and characteristics of each family of capacitor
- Why electrolytic capacitors are polarity sensitive
- The basics of variable capacitors

4.1: FIXED CAPACITORS

Capacitors are the second most common of all discrete devices in electronic circuits. Early capacitors were termed "condensers," and the term is occasionally still encountered in the literature. The earliest capacitor was the Leyden jar, developed in the city of Leyden in the Netherlands, and used to store electrostatic charge.

There are many types of capacitors, but they can be grouped into five main classifications.

1. glass, mica and porcelain
2. multi-layer ceramic
3. plastic
4. electrolytic
5. specialty

4.1.1: Capacitor Operation and Functions

As in the Leyden jar, the purpose of a capacitor is to store electrostatic charge, the same as when we rub our feet on the carpet and our body becomes the storage place of the charge we have picked up. Any two conductors separated by an insulator, or dielectric, constitute a capacitor. The simplest diagram for a capacitor is the parallel-plate diagram in Figure 4.1. As a voltage causes charge to build up on one of the plates, charges of the same polarity (being repelled) are pushed away from the opposite plate, giving it the opposite charge. This creates an electrostatic field, which stores energy.

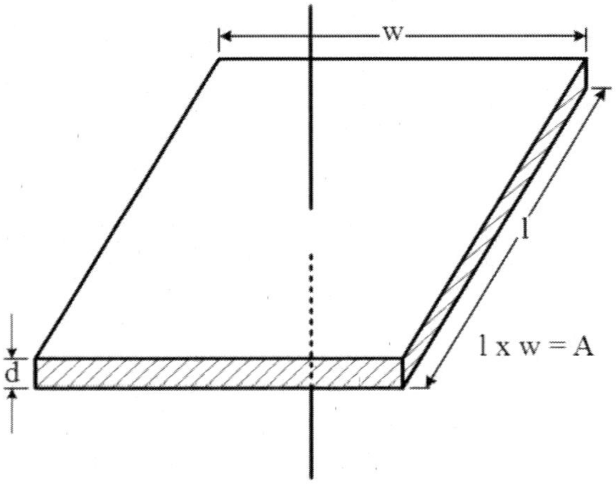

Figure 4.1 A model for a simple parallel-place capacitor. d = distance between places, l = length of plate, and w = width of plate.

The process of building up charge in a capacitor is very much analogous to the process of building up air pressure in a two-ported chamber separated by an air-tight rubber diaphragm, as shown in Figure 4.2. As air is introduced into the left side of the chamber, the diaphragm displaces to the right,

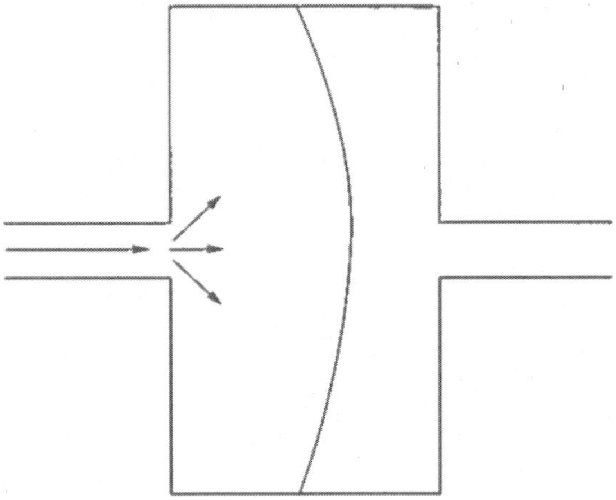

Figure 4.2 Two-ported reservoir analog for a capacitor.

pushing out air. As the air pressure is increased, the distention of the diaphragm increases, allowing more air into the left side of the chamber while at the same time pushing a similar quantity of air out the right side. If the chamber depicted in Figure 4.3 were capped on both ends and removed from the page,

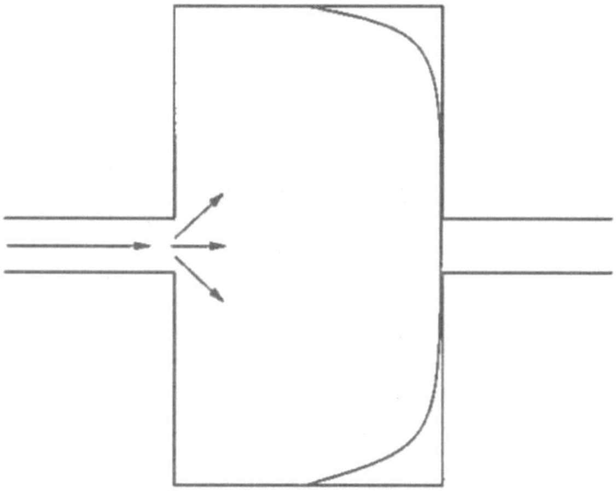

Figure 4.3 Two-ported reservoir filled with air.

it would be a small, self-contained source of air under pressure. The same is true for capacitors, as described in the preceding paragraph. Once charged and then disconnected from the circuit, a capacitor becomes a small, self-contained source of electricity.

The capacitance of a given capacitor, as already discussed, is given by the formula:

$$C = \frac{\varepsilon_0 \varepsilon_r A}{d} \qquad \text{Eqn 4.1}$$

where C is capacitance in Farads, A is the overlapping area of the capacitor plates, d is the distance between the plates, ε_o is the absolute permittivity of free space (8.85x10-12 F/m) and ε_r (epsilon relative) is the dielectric constant.

This equation is actually quite easy to understand. If we take a material with a dielectric constant (ε_r) of 1.0, such as air, add two plates each with an area of 1 m^2, and separate them by 1 m (d), we would have a capacitance of:

$$C = \frac{(8.85\ X\ 10^{-12})(1.0)(1.0\ m^2)}{1m} = 8.85\ X\ 10^{-12}, or\ 8.85\ pF$$

In the previous chapter on resistors, we were concerned about minimizing the capacitance between adjacent conductors. However, in this application, we want to have capacitance, and so our focus turns to how to create the desired capacitance appropriately, with minimal parasitics (resistance and inductance).

4.1.2: Capacitor Characteristics

Once a capacitor is charged, it would be ideal if that charge would stay there indefinitely. However, as there is no such thing as a perfect insulator, the charges eventually work their way through the insulating material and the capacitor self-discharges. This is known as *leakage*. Good capacitors, therefore, would have low leakage.

It is possible to build up charge on the plates of the capacitor such that eventually the voltage exceeds the insulator's ability to withstand it, and the dielectric breaks down. The classic form of this dielectric breakdown is seen when the air, (a dielectric) builds up a charge of several million Volts, and breaks down as a discharge of lightning passes through it. Once the discharge is over, the air fills in the place where the discharge occurred and the air dielectric heals itself. However, self-healing is not possible with solid materials such as those commonly used in capacitors. Once the dielectric *breakdown voltage* is exceeded and the small discharge occurs inside a capacitor, the capacitor is permanently ruined. Thus, one desirable characteristic of a good capacitor is a high *dielectric strength*.

It should also be mentioned that this breakdown voltage is proportional to the thickness of the dielectric, and indeed is usually given in units of V/mil. Thus, a capacitor with a dielectric 2 mils thick would have a dielectric strength twice that of a capacitor with a dielectric only 1 mil thick, assuming identical materials. The dielectric strength of several electronic materials is given in Table 4.1.

This means that to have a physically small capacitor, one must trade off a large capacitance rating with a high voltage rating, because both require a capacitor to be physically larger. Consider, for example, the capacitors shown in Figure 4.4; they are both aluminum electrolytic capacitors of basically identical

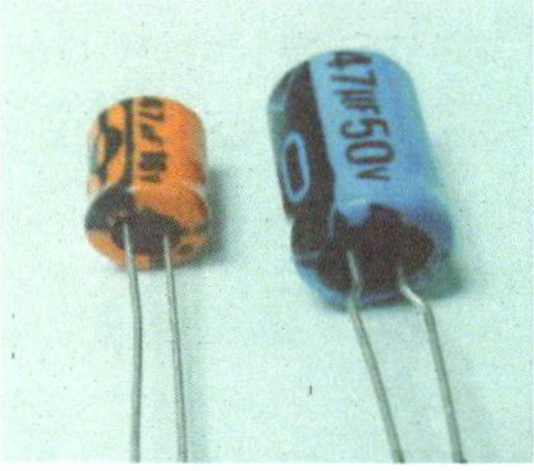

Figure 4.4 An example of the difference that the voltage rating makes. These two aluminum electrolytic capactiors both have the same capacitance rating (47µF) but have different voltage ratings; left = 16 Volts; right = 50 Volts

construction; the left one is rated at 47 µF, 16 V, while the right one is rated at 47 µF, 50 V. Only the difference in their voltage range accounts for the difference in their physical size, and this voltage rating is directly related to the dielectric strength of the dielectric material.

Dielectric Material	Dielectric Strength (V/mil) @ 60 Hz	Dielectric Constant @ 60 Hz	Dissipation Factor @ 1 MHz
Air	30-300	1.0006	
Aluminum Oxide (Alumina; Al_2O_3)	300	9.34	0.001
Ceramic (high-k)	≈200	8000	0.002
Ceramic (low-k)	≈200	6.0	0.002
FR4	360	4.2	0.01
Glass	1532	7.5	0.001
Mica	712	5.0	0.0005
Oil / Paper	≈500	4.0	
Paper	≈50	2.5	
Phenolic	400	12.0	0.8
Polycarbonate	410	2.97	0.0003 @ 60 Hz
Polyester	300	4.5	0.05
Polyimide	6500	3.4	0.01
Polypropylene	550	2.4	0.002 @ 60 Hz
Polystyrene	3174	2.6	0.0003 @ 60 Hz
Porcelain	232	≈5	
Rubber	520	3.0	0.05
Silicon Oxide (SiO_2)	≈100	4.42	0.0001
Silicon Nitride (Si_3N_4)	≈100	8.1	
Tantalum Oxide (Tantala; Ta_2O_5)	≈50	77.0	
Teflon	850 (@60 Hz)	2.0	0.0002
Vacuum (Practical)	≈10,000	1.000	0.00001

Table 4.1 Dielectric strength and dielectric constant (k or ε_r) for several electronic materials.

Ideally, the process of building up a charge in a capacitor would be without losses, such that the energy stored in a capacitor would be exactly equal to the energy returned. However, practical materials exhibit some opposition to the storage of charge; this opposition is manifested as a dissipation of energy in the form of a very small amount of heat. These losses are not significant in most cases, but are real. These losses are described in the term *dissipation factor*, which is the ratio of the reactance to the resistance of a capacitor at a given frequency. An ideal capacitor would have an infinite resistance, which would give a dissipation factor of 0. The dissipation factor of several electronic materials is given in the far right column of

Table 4.1.

Capacitors, like resistors, are desired in a very wide range of values. Typical discrete capacitor values range from 1 pF to 10,000 μF (.01F). It would be nice if every type of capacitor were available in such a wide range, but this is not practical. Table 4.2 lists the value range for several capacitor types.

As in the case for resistors, it would be ideal if all 1.0 μF capacitors were exactly 1.0 μF. We want a very tight tolerance in their values. Capacitors, in general, are not as tight in their tolerance as are resistors; typical tolerances are from 1 percent to 20 percent. A summary of the tolerances available for several capacitor types is given in Table 4.2

Capacitors should also possess very little parasitic inductance and resistance. Parasitic inductance shows up as a limitation in the frequency range of the capacitor; as the frequencies get higher, the inductance becomes dominant, and the capacitor no longer behaves as a capacitor. When this occurs, the capacitor is said to be at self-resonance, or the point where the inductive reactance and the capacitive reactance are equal.

Parasitic resistance in capacitors is known as *equivalent series resistance*, or *ESR*, and can be a serious limiting factor in the performance of large storage capacitors by interfering with the flow of peak currents in and out of the capacitor. It also causes the capacitors to not work as well in filter applications. Ratings of the ESR of several capacitor types are found in Table 4.2.

We would also like our 1.0 μF capacitor to be 1.0 μF at every temperature at which it is operated. That is, we would like it to have an excellent *temperature stability*. Several capacitor types are rated for this factor in Table 4.2.

When a capacitor is discharged, ideally all of the charge which was stored in the capacitor is available to be discharged. However, this is not always the case. In some capacitors, there is a significant amount of *dielectric absorption*, which causes some of the charge to be absorbed into the dielectric. This shows up as a residual charge in a capacitor after it has been shorted (discharged) for a time, and is an undesirable characteristic. This is particularly important in precision applications.

It should be mentioned here that the plates of a capacitor are not nearly as significant in the performance of the capacitor as the dielectric material. In all the above desired capacitor characteristics, all but ESR are determined almost entirely by the dielectric material. For this reason, capacitors types are specified by the material they use for the dielectric, and not the plate material.

4.1.3: Glass, Mica and Porcelain

These capacitor types are grouped together because the dielectric materials are all largely vitreous (glass-like) silica, and have very similar electrical characteristics, as seen in Table 4.1. In general, they are excellent for low-leakage, high-frequency applications, have very good tolerance and temperature stability, and are low in cost. Their main disadvantage is that there is a very limited range of values available, usually only up to 0.1 μF.

These capacitors are typically made from a thin disc of the dielectric material which is then coated with a layer of metal on each side. The leads are soldered to this metal, and the assembly is coated with epoxy paint. Mica capacitors are usually coated with silver, which is why they are commonly referred to as silvered mica capacitors. Figure 4.5 shows a few members of this family of capacitors. A picture and a cross-section of a silvered mica capacitor is shown in Figure 4.6.

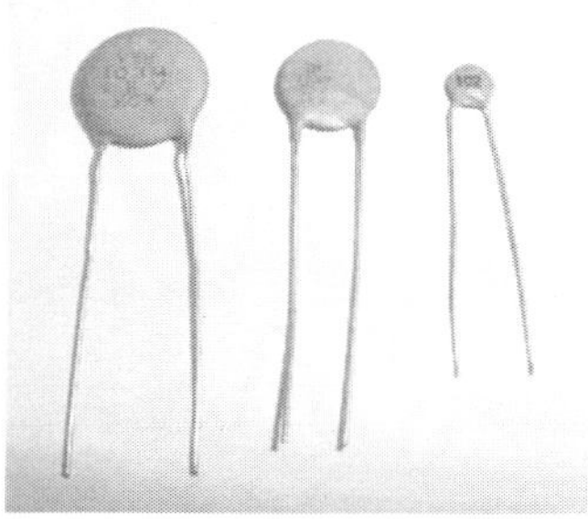

Figure 4.5 Example of several silvered mica capacitors.

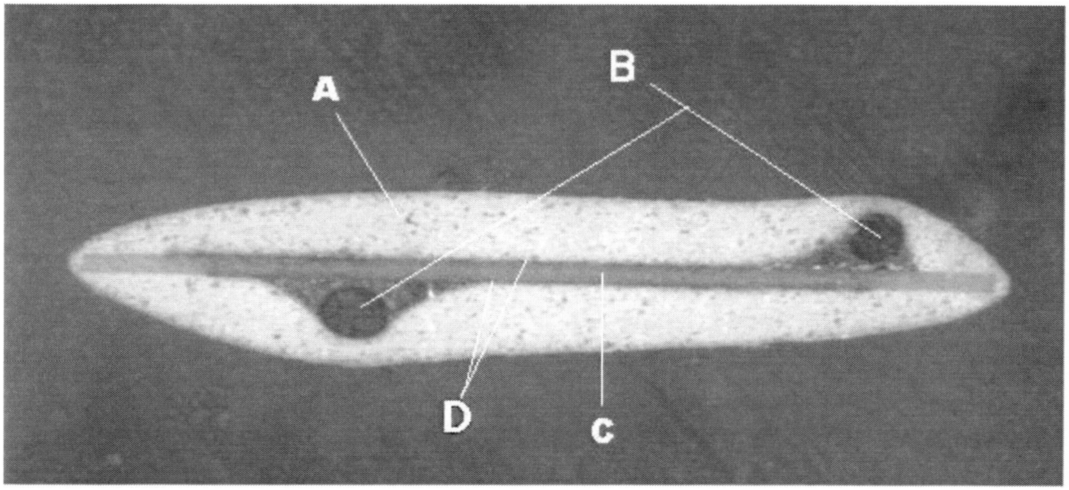

Figure 4.6 A cross-section of a silvered mica capacitor. (a) epoxy coating, (b) soldered leads, (c) sheet of mica, (d) thin layer of silver on each side of the mica sheet.

Type	Value Range	Tolerance (±%)	Dielectric Absorption	Maximum Voltage Range (V)	Useful Frequency Range (Hz)	Temperature Stability	Leakage	Cost	ESR	Comments
Glass	10pF-1nF	5 - 20	Low	100 - 600	$0 - 10^9$	Good	Excellent	Low	Very low	Excellent long-term stability
Porcelain	50pF-.1μF	5 - 20	Low	50 - 400	$0 - 10^8$	Good	Good	Low	Very low	Good long-term stability
Silvered Mica	1pF-.1μF	1 - 20	Low	500V-75kV	$10^3 - 10^{10}$	Good	Good	Low	Very low	Excellent; great for RF
M-L Ceramic (high-k)	50pf-10μF	+10 to -20	Low	up to 6000	$10^3 - 10^8$	Poor	Poor	Low	Very low	Small, inexpensive, popular
M-L Ceramic (low-k)	1pF-10nF	2 - 20	Very low	up to 6000	$10^3 - 10^{10}$	Excellent	Good	Low	Very low	Several tempco's
Polycarbonate	50pF-30μF	1 - 5	Very low	50 - 800	$0 - 10^8$	Excellent	Good	Low	Very low	High quality, small
Polyester (Mylar)	1nF-50μF	5 - 20	Very low	50 - 600	$10^2 - 10^8$	Poor	Good	Low	Very low	Inexpensive, popular
Polypropylene	50pF-50μF	1 - 5	Very low	50 - 800	$0 - 10^{10}$	Good	Excellent	Low	Very low	Low dielectric absorption
Polystyrene	10pF-10μF	.5 - 5	Very low	up to 1000	$0 - 10^{10}$	Good	Excellent	Low	Very low	Low dielectric absorption
Teflon (PTFE)	1nF-10μF	1 - 5	Very low	50 - 200	$0 - 10^{10}$	Best	Best	Low	Very low	Lowest dielectric absorption
Aluminum (electrolytic)	.1μF-.05F	100 to -20	Medium	3 - 500	$0 - 10^4$	Terrible	Awful	Medium	High	Low cost, high capacitance
Tantalum (electrolytic)	.1μF-500μF	10 - 50	Medium	1 - 100	$0 - 10^5$	Poor	Bad	High	Medium	High volumetric efficiency
Double-etched aluminum electrolytic	.1F-100F	20 - 40	Medium	1.5 - 6	$0 - 10^3$	Terrible	Awful	Very high	High	Highest volumetric efficiency
Air - variable	10pF-500pF	.5 - 5	Very low	500	$10^2 - 10^8$	Good	Excellent	High	Low	RF tuners, oscillators
Paper/Oil	1nF-50μF	10 - 20	Medium	100-100kV	$10^2 - 10^8$	Moderate	Good	Very high	Very low	High energy density
Vacuum	1pF-5nF	10 - 20	Extremely low	2kV-100kV	$10^1 - 10^8$	Good	Excellent	High	Low	Power transmission

Table 4.2 Characteristics of various capacitor types.

4.1.4: Multi-layer Ceramic

Multi-layer ceramic capacitors have become overwhelmingly popular in general-purpose surface-mount applications. They are also very popular for many through-hole applications. This is because they share all the advantageous characteristics of the glass, mica and porcelain family (ceramics are actually in the same vitreous family), but are available in a wider range of values and have an improved temperature stability. Several types of ceramic material have been developed specifically for application in capacitors, and it is these that make up this family of capacitors.

As can be seen in Tables 4.1 and 4.2, there are two main categories of multi-layer ceramic capacitors, high-k and low-k. The "k" refers to the dielectric constant of the dielectric material (another term for ε_r). Obviously, high-k capacitors have a higher capacitance value range, since capacitance is directly proportional to the dielectric constant. The disadvantage of high-k capacitors is their poorer temperature stability. The low-k ceramic capacitors have much better temperature stability. Table 4.3 summarizes four popular types of multi-layer ceramic capacitors and their value ranges, temperature ranges and temperature stabilities. Note that as the capacitance goes up, the temperature stability goes down.

Type	Value Range	Temperature Range	Temperature Stability
NPO	1 - 1000 pF	-55° to +125°C	Stable (±0%)
X7R	1000 pF - .1 µF	-55° to +125°C	±15%
Z5U	.1 µF - .4 µF	+10° to +85°C	+22%, -56%
Y5V	.2 µF - 1.0 µF	-25° to +85°C	+22%, -82%

Table 4.3 Characteristics of multi-layer capacitors.

Figure 4.7 shows some examples of multi-layer ceramic capacitors. A cross-section of a multi-layer ceramic capacitor is shown in Figure 4.8. Using these multiple layers and the very thin dielectric, significant capacitances can be achieved, up to about 1.0 µF. This is compared to the disk structure of the glass, mica and porcelain family, which typically only achieves values of 1 nF without multiple layers.

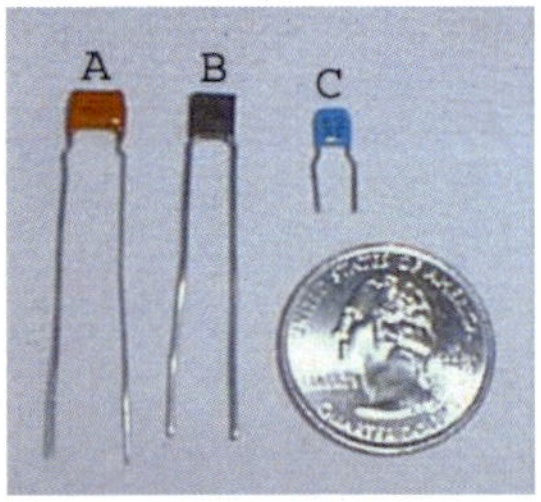

Figure 4.7 An example of three multi-layer ceramic capacitors. (a) 1.0 μF, (b) 0.1 μF, (c) 22nF.

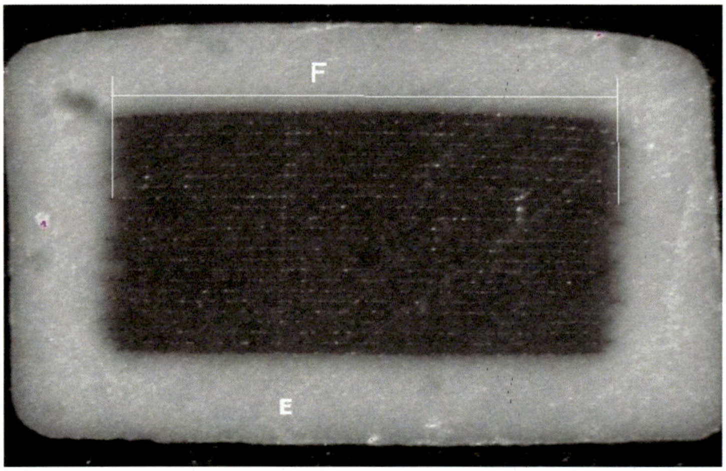

Figure 4.8 A cross section of a multilayer ceramic capacitor. (E) epoxy coating, (F) overlapping layers of very thin ceramic sheets.

4.1.5: Plastic

Plastic capacitors, in general, have not risen in popularity, but due to some of their remarkable properties, they still enjoy significant usage in some applications. As can be seen in Table 4.2, their characteristics include a wide value range (up to about 50 μF), excellent tolerances, very low leakage, excellent frequency range, good to excellent temperature stability, low cost, and very low dielectric absorption. The main reason they have become less popular has been the rising popularity of the multi-layer ceramic capacitor for surface mount applications.

Figure 4.9 shows some common plastic capacitors. As shown in Figure 4.10, plastic capacitors are constructed from alternating layers of dielectric and foil, with the foil layers aligned so that they extend on opposite ends. These opposite ends are where the end caps and leads are attached, after which the assembly is coated in a final layer of plastic.

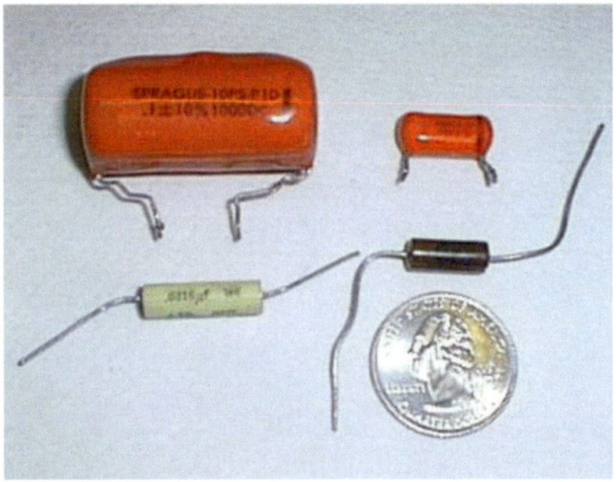

Figure 4.9 Several plastic capacitors.

Figure 4.10 Construction of a plastic capacitor. *From Introductory DC/AC Circuits by Cook, © 1999. Reprinted by permission of Pearson Education, Inc. Upper Saddle River, NJ 07458.*

4.1.6: Electrolytic

Electrolytic capacitors are, in general, the poorest performers of all the capacitor groups. As seen in Table 4.2, their tolerances are poor to awful, their dielectric absorption is not good, their useful frequency range is poor, their temperature stability is ghastly, and their leakage is awful. In spite of all these disadvantages, almost all electronic circuits have at least one, and usually several. The reason for their popularity is their volumetric efficiency, which simply means that they provide a large amount of capacitance in a small amount of space (volume). Notice that their capacitance range starts where most of the other families end, at around 1 μF, and goes clear up to 100 F. Figure 4.11 shows some examples of these capacitor types.

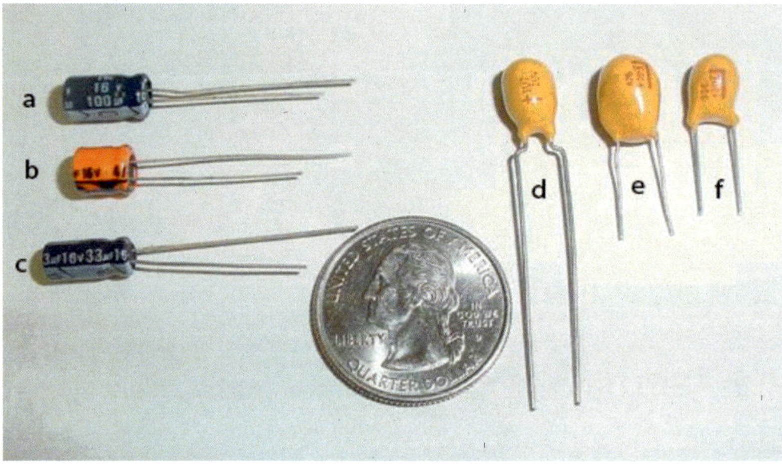

Figure 4.11 Aluminum and tantalum electrolytic capacitors. (a) 100 μF, 50 V; (b) 47 μF, 50 V; (c) 33μF, 16 V; (d) 100 μF, 16V; (e) 47 μF, 16 V (f) 33 μF, 16 V. Capacitors a, b, and c are aluminum. Capacitors d, e, and f are tantalum electrolytics.

The simple fact is that if you need a large amount of capacitance (>50 μF), your only option is to go with electrolytic capacitors. So how can capacitors that perform so poorly be tolerated in electronic circuits? The answer lies in the ingenuity of design engineers, and the many ways that have been devised to circumvent this problem. This is evident from the fact that so many electronic circuits use electrolytic capacitors.

Figure 4.12 depicts how aluminum electrolytic capacitors are constructed before the dielectric layer is grown. The electrolyte (a conductive liquid or paste) between the layers of foil serves to allow electricity to pass through all the layers of the capacitor. After the capacitor is constructed, electricity is applied to the terminals, and a layer of aluminum oxide (alumina) begins to grow on one side of the capacitor. When this oxide layer is complete, electricity no longer flows and the capacitor dielectric is ready. At this point, most of the excess electrolyte is removed (although some still remains), and the capacitor is encased in an aluminum can with a characteristic crimp on one end. This capacitor is also labeled with an indicator as to its negative terminal, because it is now a polarized capacitor. It is polarized because of the growth process of the oxide. If a voltage is applied in the opposite polarity, the growth process begins to reverse, and the oxide layer soon diminishes and breaks down, usually causing a catastrophic failure (explosion) of the capacitor.

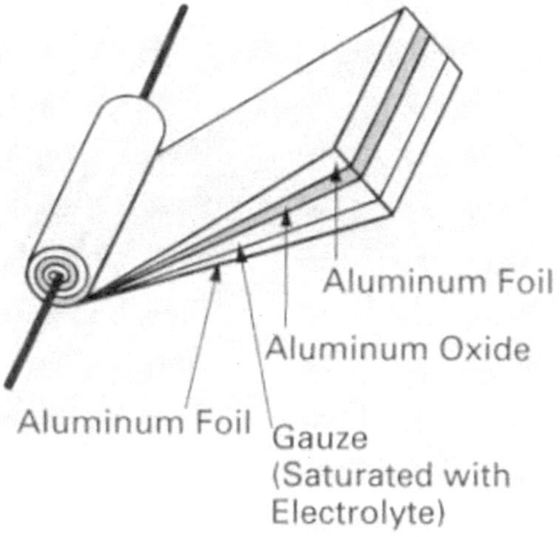

Figure 4.12 Construction of an aluminum electrolytic capacitor. *From Introductory DC/AC Circuits by Cook, © 1999. Reprinted by permission of Pearson Education, Inc. Upper Saddle River, NJ 07458.*

The oxide layer of an aluminum electrolytic capacitor is very thin, much thinner than the dielectric can be made with plastic or ceramic capacitors. Because capacitance is inversely proportional to the distance between the plates, and these have a very thin dielectric, it follows that this capacitance can be significant. It is this main factor (the thin dielectric) that accounts for the high volumetric efficiency of electrolytic capacitors.

Figure 4.13 depicts how tantalum electrolytic capacitors are constructed. Because of the extremely porous nature of the tantalum pellet, there is a large surface area. It is also evident that the tantalum oxide (tantala) dielectric is very thin. These two factors account for much of the high volumetric efficiency of tantalum capacitors; it is roughly equivalent to that of aluminum electrolytics. The final factor that accounts for the volumetric efficiency of tantalum electrolytics is that the dielectric constant of tantala is so much higher than that of alumina (77 versus 9.3).

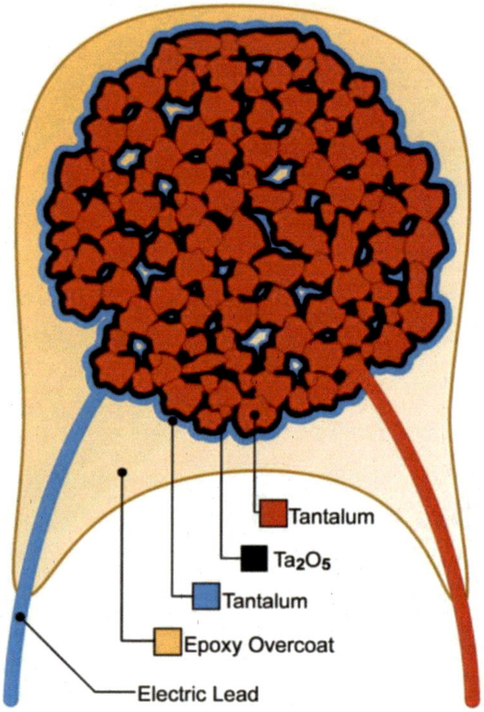

Figure 4.13 Construction of a tantalum electrolytic capacitor. *Courtesy of Richard Fry.*

Most tantalum electrolytic capacitors are slightly smaller than a comparable aluminum electrolytic; a comparison of equivalent tantalum and aluminum capacitors can be seen in Figure 4.14. The main advantage of tantalum electrolytics is their higher performance (lower leakage, better temperature stability, higher frequency range), as compared to aluminum electrolytics. Because of this, they would certainly be more popular than aluminum electrolytics, if it were not for the fact that tantalum electrolytics are more expensive than aluminum electrolytics.

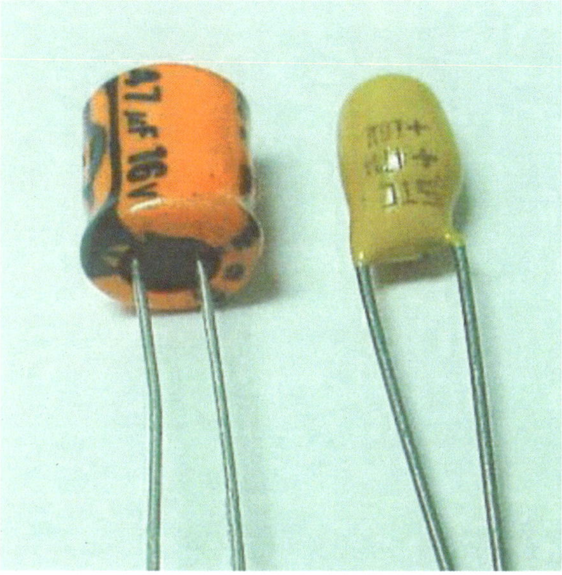

Figure 4.14 Comparison of aluminum and tantalum electrolytic capacitors with the same capacitance and voltage ratings (47 µF, 16 V).

The final member of the family of electrolytic capacitors is double-etched electrolytics. These are aluminum electrolytic capacitors which have been treated by a process of coarse etching of the plates, followed by fine etching, as depicted in Figure 4.15. This double-etching process leaves a plate with a heavily pitted surface, greatly increasing its surface area. On this surface is then grown the layer of aluminum oxide to form the dielectric. Thus, according to Equation 4.1, this greatly increased surface

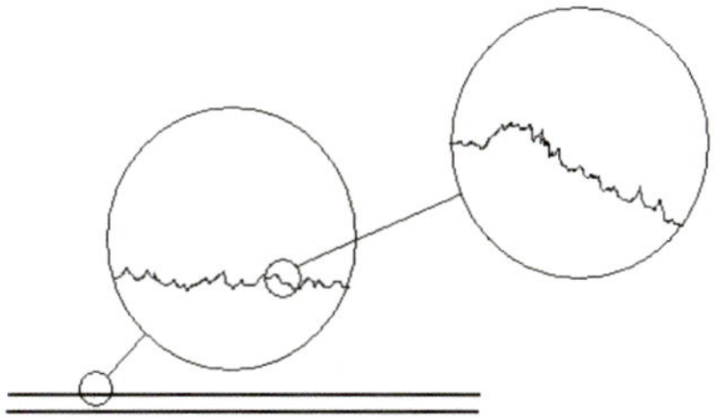

Figure 4.15 Successive close-up views of the foil of a double-etched electrolytic capacitor. The greatly increased surface area provides for a dramatic increase in capacitance.

area equates to a large increase in capacitance. The layers are also very close (d of Equation 4.1 is very small), which also leads to increased capacitance, as well as a very low voltage rating. Figure 4.16 shows some double-etched electrolytic capacitors, along with their extraordinarily large capacitance ratings and low voltage ratings. It is truly a manufacturing marvel that such large capacitances are now available in such small sizes.

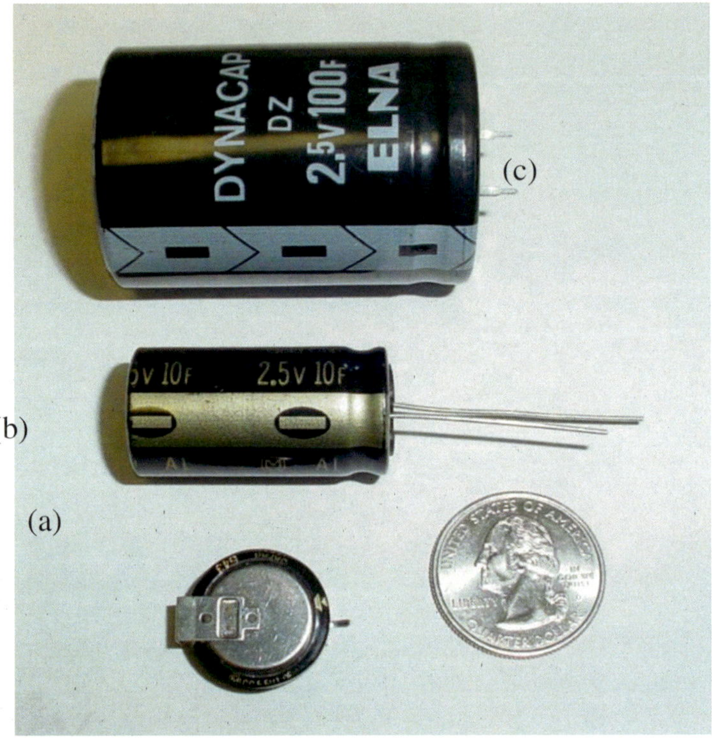

Figure 4.16 Several double-etched electrolytic "super" capacitors. (a) 1.0 F, 5.5 V (b) 3.3 F, 5.5 V, (c) 10 F, 2.5 V

4.1.7: Packages for Printed Wiring Board Capacitors

Capacitor package types are many, each tailored to the physical properties of the dielectric from which the capacitor is made. Figure 4.17 shows several of these package types. The axial and radial package (through-hole) types are diminishing in popularity while the surface mount package types are increasing in popularity. However, the selection of capacitor types in surface mount packages is much more restricted, generally being confined to only multi-layer ceramics and electrolytics. Most surface-mount capacitor packages (chip capacitors) are designated by their physical X-Y dimensions, as are surface-mount resistor packages (see chapter 3), but often are much thicker than surface-mount resistors. They also come in much larger packages than the largest surface-mount resistor package.

Aluminum electrolytic capacitors also come in miniature cans on top of a surface-mount carrier, especially for the larger values. An example of this is seen in (l) of Figure 4.17.

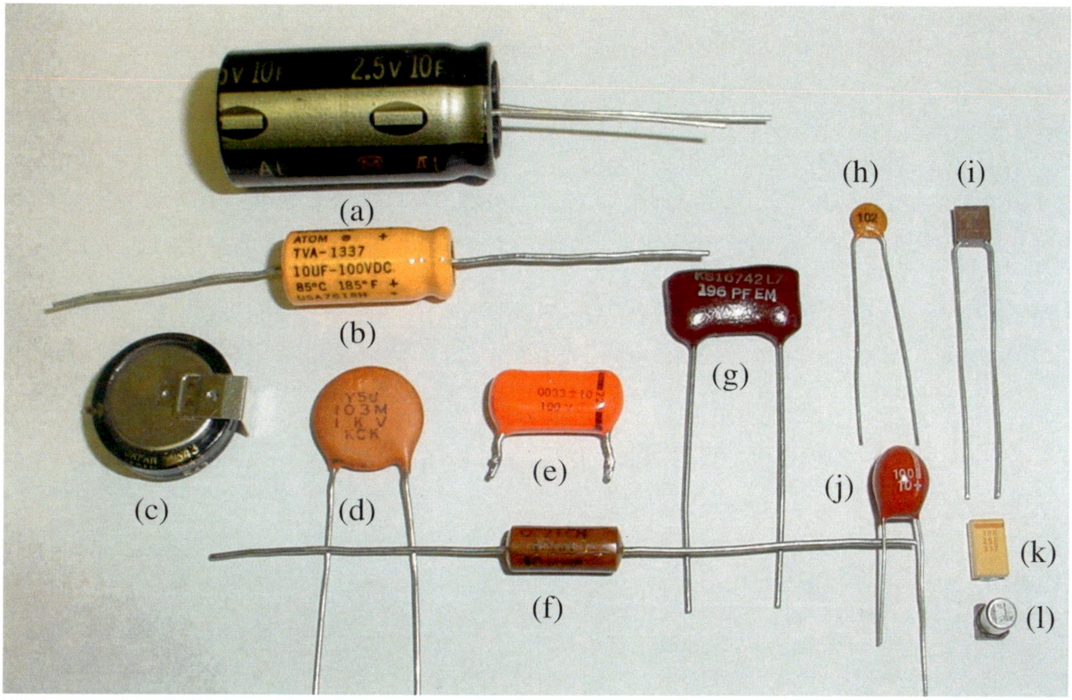

Figure 4.17 Several different capacitor types. (a) radial super capacitor, (b) axial aluminum electrolytic, (c) solder-tab super capacitor, (d) radial ceramic, (e) radial plastic, (f) axial plastic, (g) radial ceramic, (h) radial silvered mica, (i) radial multi-layer ceramic, (j) radial tantalum electrolytic, (k) rectangular surface-mount electrolytic, and (l) canister surface-mount electrolytic.

4.1.8: Specialty: Paper/Oil and Vacuum

The energy stored in a capacitor is given by the equation, $W = QV^2$. If one wishes to store a large amount of energy in a capacitor, clearly it is better to increase its voltage rating than its charge storage (Q, or basically capacitance). For this reason, and a few others, all capacitors that are used to store large amounts of energy usually have very high voltage ratings (>1kV). Such capacitors are used in applications where large doses of energy are needed in very short bursts, such as in defibrillators, firing rail guns, and energy plasma/fusion research. To meet this need for high voltage ratings, paper/oil and vacuum capacitors have been developed.

In paper/oil capacitors, the plates are separated by a thin layer of paper which is impregnated with oil, and the entire assembly is usually sealed with oil inside to keep the plates separated electrically. Oil has a higher dielectric strength than air, and therefore can provide higher voltage ratings. Such capacitors are usually quite large and expensive, and are not suited for mounting on circuit boards or in most electronic equipment.

Vacuum capacitors are even more expensive, but have the advantage of even higher voltage ratings. If we could achieve and maintain a perfect vacuum, it would have an infinite voltage rating. Dielectric breakdown requires molecules to be present, since it is the ionization of molecules that produces the eventual path for the arc to form. If there are no molecules present (a perfect vacuum), there is nothing to ionize to form an arc. While we cannot achieve a perfect vacuum, the very low vacuums we can achieve allow us to make vacuum capacitors with ratings exceeding 100 kV, providing very large amounts of energy storage. As with paper/oil capacitors, they are not suited for mounting on circuit boards or in most electronic equipment.

Most people who work in electronics will probably not come across either of these specialty capacitor types. Figures 4.18 and 4.19 show an example of each.

Figure 4.18 High voltage, oil filled capacitors for high energy applications. *Photo courtesy of General Atomics Energy Products, www.hvpower.com.*

Figure 4.19 High voltage vacuum capacitor for high energy applications. *Courtesy of American Trans-Coil Corporation, http://www.atc.us.com/vacap.html.*

4.1.9: Integrated Circuit Capacitors

Much has been said in the preceding sections about discrete capacitors, but until now, integrated circuit capacitors have been ignored. This is because practical capacitors are not very feasible to make on ICs, for two main reasons:

1. they use a lot of real estate, and therefore are expensive (relative to the transistors).
2. the easily formed dielectric available (SiO2) does not have a very high dielectric constant (4.42); large-value capacitors are not available.

However, all this does not mean that ICs do not have capacitors; indeed, in DRAM ICs there are undoubtedly more capacitors than there are discrete capacitors in all the rest of electronics combined.

ICs have capacitors of two different types: those for DRAM cells, and those for general purpose applications. DRAM (Dynamic Random-Access Memory) stores data in the form of charge on a very small capacitor, gated by a single transistor, as shown in Figure 4.20. The capacitor is formed in a trench process, using the vertical dimension to add surface area, rather than the very "expensive" planar dimension ("expensive" meaning it takes up valuable real estate). Several such capacitors are shown in cross-section in Figure 4.21. These capacitors are made by the trillions, one for each bit; a computer with 16 Gbytes of memory would therefore have over (8 x 16 billion) ≈ 128 billion! These capacitors are vanishingly small, with typical values of 15 fF (f = femto = 10^{-15}). Silicon nitride has replaced silicon oxide as the dielectric, because of its higher dielectric constant of 8.1.

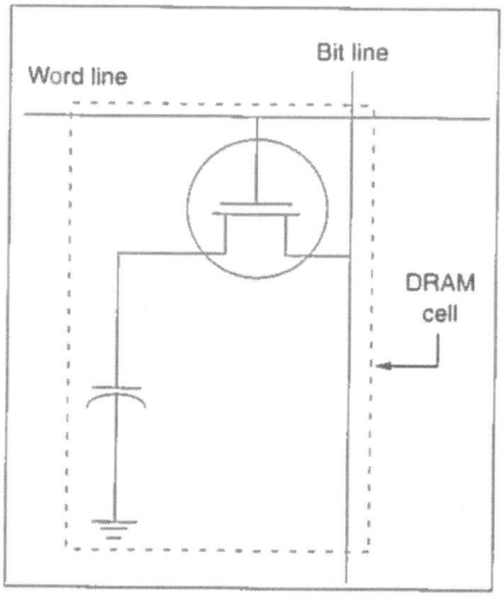

Figure 4.20 A DRAM storage cell, including one metal-oxide-semiconductor field-effect transistor (MOSFET) and one capacitor.

Figure 4.21 A cross section of a storage cell for 16 Mbit DRAM. From IBM Microelectronics Intranet. Courtesy of IBM Corporation.

The other type of IC capacitor is a general purpose capacitor. The values available are very small, ranging from tens of fF to about 50 pF. The characteristics of IC capacitors are very similar to glass discrete capacitors, since most glass is basically SiO_2.

4.2: VARIABLE CAPACITORS

Variable capacitors have historically been only one major type: air-variable, meaning they use air as the dielectric. They are constructed from parallel plates, one set of which can be rotated in and out of the fixed set, as shown in Figure 4.22. The large kind shown here used to be very common in tuned equipment, such as radios and television; the tuning dial was directly or indirectly linked mechanically to the variable capacitor. However, modern electronics uses digital tuners, which do not use variable capacitors, so these large variable capacitors have virtually disappeared in new equipment.

Figure 4.22 Large, air-variable capacitors.

There are still a few applications where a frequency, delay or pulse width needs to be set in the factory, or where experimentation calls for a small variable capacitance. This need is met by trimmer capacitors, which are similar in function to trimmer potentiometers in that they are not adjusted by the end user of the product, but are only intended to be adjusted by the manufacturer. Figure 4.23 shows an example of a trimmer capacitor. Typical values for trimmer capacitors only go up to about 50 pF.

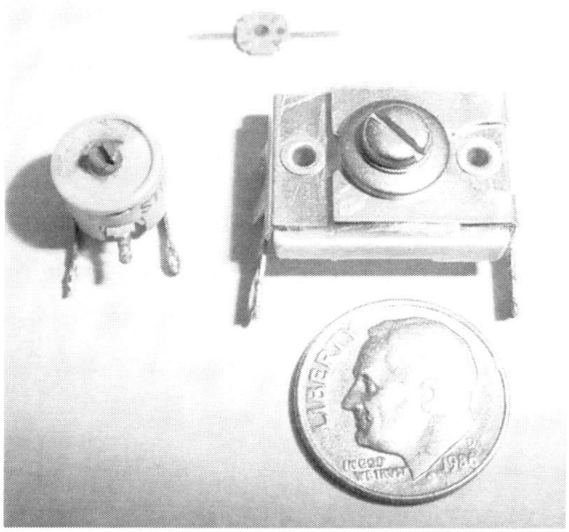

Figure 4.23 Trimmer capacitors.

CHAPTER SUMMARY

Capacitors are the third most common component in electronic circuits, behind the transistor and the resistor. There are many desirable characteristics for an ideal capacitor, including low leakage, high dielectric strength, low dissipation factor, high dielectric constant, wide value range, low tolerance, low cost, low ESR, high temperature stability, low dielectric absorption, and a wide useful frequency range. The performance of the main types of capacitors is summarized in Table 4.2.

Capacitors within each of the five families of capacitors have similar construction and characteristics. Understanding how a capacitor is constructed is helpful in understanding its performance in each of the desirable characteristics. No single type of capacitor is best in all characteristics.

Variable capacitors are also available, both as trimmers and as adjustable capacitors, but these are much less common than variable resistors. Values for variable capacitors are small, ranging only up to about 50 pF.

FOCUS QUESTIONS

1. What are the five families of capacitor types?
2. What are 10 of the desirable characteristics of a capacitor?
3. Find the capacitance of a polyester capacitor with an area of 20 mm^2 and a dielectric thickness of 0.1 mm.
4. Find the area of a low-k ceramic capacitor with a thickness of 0.085 mm and a capacitance of 10 nF.
5. What is the relationship between dielectric strength and dielectric constant?
6. What does it mean if a given capacitor has a high dissipation factor?
7. Which of the four capacitor types in Table 4.3 would you expect to be made of low-k ceramic dielectric? What is the main advantage of this type of capacitor?
8. What are two of the main strengths of the glass, mica and porcelain family of capacitors?
9. What are two of the main strengths of the plastic family of capacitors?
10. What is the main (and almost only) strength of the electrolytic family of capacitors?
11. What breakthrough would have to occur in order to make a non-electrolytic capacitor with a capacitance of 1.0 F?
12. What is the main advantage of paper/oil and vacuum capacitors? Why would you not expect to find them in most electronic equipment?

FOR ADDITIONAL READING

Conway, B. E., "Electrochemical Supercapacitors: Scientific Fundamentals and Technological Applications", Kluwer Academic/Plenum Press, New York, NY, 1999.
Fink, Donald G. and Christiansen, Donald, "Electronics Engineer's Handbook, Section 7, McGraw-Hill, New York, NY, 1989.
Livingston, James D., "Electronic Properties of Engineering Materials", John Wiley & Sons, New York, NY, 1999.
Munshi, M. Zafar A., "Handbook of Solid State Batteries and Capacitors", World Scientific Pub., River Edge, NJ, 1995.
Sinclair, Ian Robertson, "Passive Components: A User's Guide", Oxford, 1990.
Wolf, Stanley, and R.F.M. Smith, "Student Reference Manual for Electronic Instrumentation Laboratories", Chapter 11, Prentice-Hall, Englewood Cliffs, NJ, 1990.

CHAPTER 5: DISCRETE TRANSISTORS AND DIODES; CONNECTORS

CHAPTER OVERVIEW

This chapter covers the following concepts:

- Discrete transistors and diodes usually have more than one package option.
- Connectors are an important part of the overall design project.
- Knowing the right material to choose for each part of a connector is essential to choosing the right connector for a given application.
- The connection process is characterized by three steps: contact, swage, and swipe.

5.1: DISCRETE TRANSISTORS AND DIODES

Transistors and diodes are typically chosen for their performance characteristics. However, after the type of transistor or diode has been chosen, there still remains the decision of the package for the device. Most discrete transistors and diodes come in two or three package styles, each with their own set of characteristics. Before we discuss the package styles available, we should first discuss what we want in a good diode or transistor package.

Ideally, the package would be completely sealed against the environment; that is, it would be an hermetic package. Such packages are made either of ceramic or metal; plastic packages do not provide hermeticity.

Next, the package must be small and light, so that it will not take up much space or add much weight. Plastic packages fit this criteria very nicely; some metal packages also do.

The final key property we would like of our transistor or diode package is that it provide an excellent path for the escape of the heat generated in the transistor itself. Plastic packages do not perform well in this category, while ceramic and metal ones do. However, there is another tradeoff here. If we want our package to dissipate large amounts of heat (>1W), it must have a large surface area, which goes against another of our wishes, small size. This is a classic engineering tradeoff, and the best compromise possible will be the best overall solution.

Figure 5.1 shows some common plastic, ceramic and metal packages available for discrete transistors and diodes; these packages are also sometimes used for simple ICs such as voltage references, voltage regulators and op amps. The DO- designation is standard for Diode Outline; the TO- designation is standard for Transistor Outline. The numbers after the DO- or TO- designation are set by JEDEC standards; publications detailing each package outline can be found at their website at www.jedec.org.

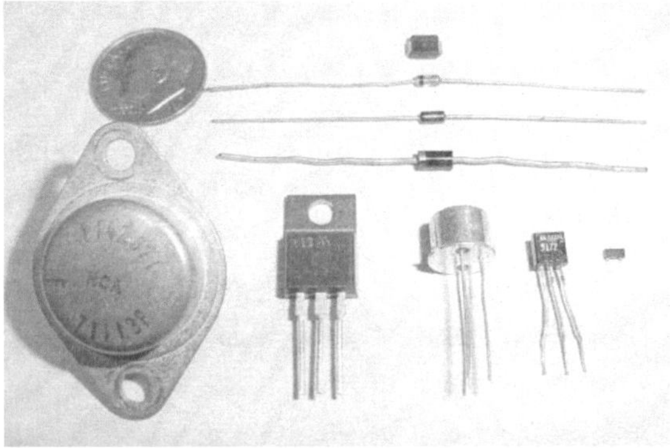

Figure 5.1. Discrete transistor and diode packages.

Most often, the key deciding factor in the package type for discrete transistors and diodes is the amount of power dissipated.

The process by which the actual diode or transistor is made is the process by which ICs are made, and will be covered in chapters 6-8.

5.2: CONNECTORS

Connectors are specific electric or electronic devices which are made to connect one part of a circuit to another. To make a working circuit, all devices must be connected, by some means, to other devices, or to the circuit inputs and outputs. Connections that are permanent or semi-permanent are made by means that will be covered in chapter 12. Connections that are separable are made with connectors; nearly every electronic product has at least one connector.

5.2.1: Connector Components and Specifications

At first blush, it may appear that there is not much to a connector - just wires connecting one point to another. But underneath the appearance of simplicity is a great deal of engineering work that has gone into making each connector work well in its intended application. A strong hint of their significance can be seen in Figure 5.2, which shows that connections and connectors account for 60 percent of the failures in electronic systems. In the author's experience, this number is even higher – probably 75%.

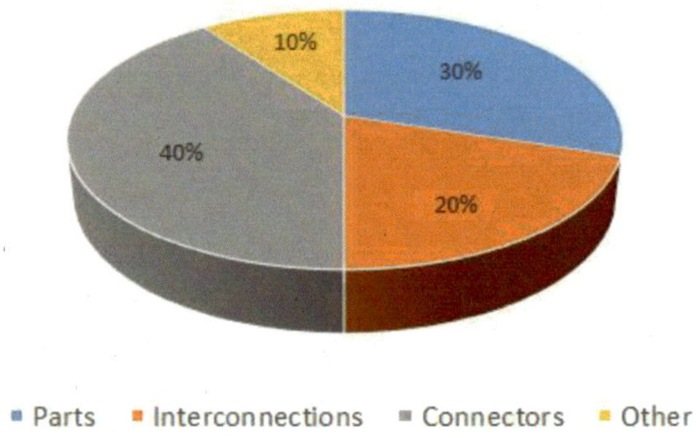

Figure 5.2. Failures in electronic systems and their causes.

Separable connectors are designed to connect one or more signals between two points in a circuit or product. They are generally designed to operate for about 1,000 mating cycles; high-use connectors are designed for up to 10,000 mating cycles. An example of a common high-use connector is an ordinary house lamp or appliance plug, and its corresponding receptacle. The plug end is usually termed the male end, while the receptacle end is termed the female end.

Connectors are made up of four main components, as shown in Figure 5.3.

1. the housing, or header
2. the pins
3. the wire
4. the wire insulation

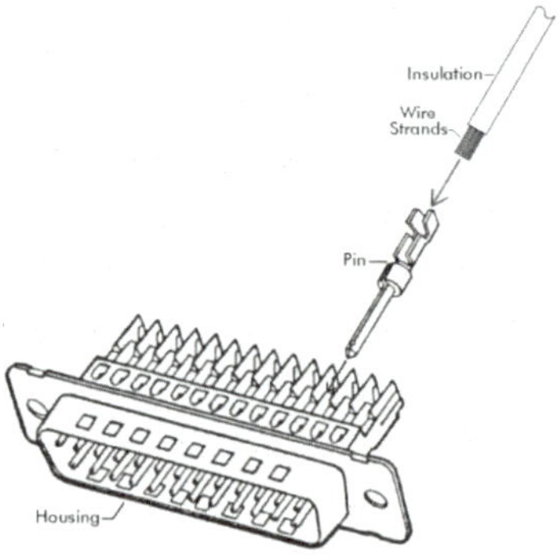

Figure 5.3. The main components of a typical connector.

Each of these components of a connector serve different functions, are made from different materials, and have different engineering specifications.
The main specifications of a connector are the following:

1. the number of pins
2. the operating temperature range
3. the voltage

4. the current
5. the number of mating cycles
6. the environmental conditions
7. the insertion force
8. the physical size

The number of pins is primarily determined by the application, and no ideal number exists. The size of the connector is usually proportional to the number of pins. However, it is usually desirable to fit all the pins in as small a space as possible.

The operating temperature range of the connector is usually specified in degrees Celsius, and commonly comes in commercial (0 to 55°C), industrial (-15 to +85°C) and military (-55 to +125°C) grades, with wider ranges possible for specialty applications. This rating is determined primarily by the material used for the housing and the wire insulation.

The voltage rating of the connector is determined primarily by these same connector components, the housing and the wire insulation. It should also be recognized that the voltage rating of a connector is generally proportional to the size, since a higher voltage needs thicker insulation or greater pin separation.

The current rating of the connector is determined primarily by the pins and the wire. Again, we must recognize that higher current ratings require larger pins and wire, so the size of a connector is also proportional to its current rating.

The number of mating cycles for a connector is determined by the materials used for the pins, and the insertion force. The environmental conditions are determined by the housing type, the housing material, and the wire insulation material. The insertion force is determined by the design of the pins.

5.2.2: Connector Operation

Ideally, every time a connector pin mates with a connector receptacle, the connection formed has no parasitic resistance. However, a number of factors combine to prevent this ideal situation in most cases. The main opposition comes from the oxygen in our atmosphere, which readily combines with most metals to form a layer of oxide (usually non-conductive) on the surface of the metal. Thus, if the pin and receptacle materials, both being metal, have a thin layer of oxide on them, the connection that results when they are brought into contact with each other will suffer from the parasitic resistance of their respective oxides. Other factors that contribute to this parasitic resistance, usually called contact resistance, include dust, moisture, wear, contact force, base materials, surface materials, and contact operation. Well-designed connectors are able to surmount these obstacles and provide reliable a connection with each mating cycle. Most such connectors rely on a three-step process in the mating cycle: contact, swage, and swipe. These are depicted in Figure 5.4

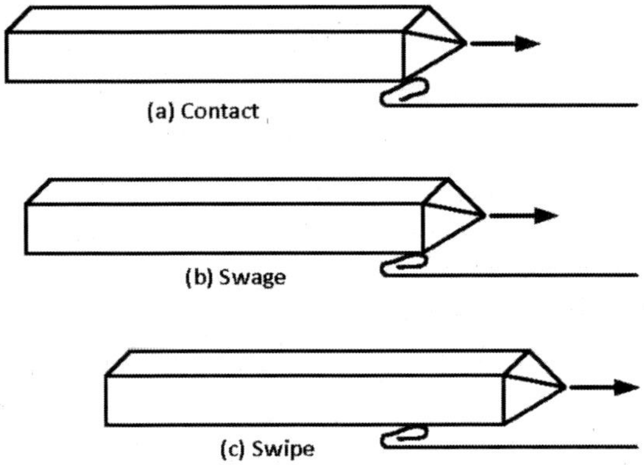

Figure 5.4. The connection process. (a) contact, (b) swage, (c) swipe.

Contact occurs when the pin and receptacle come into physical contact with each other. At this point, it is usually the thin layer of oxides on each surface that are in contact, so the electrical connection has not yet been reliably established. Swage refers to the bending or spring-compression action of the pin and receptacle after contact occurs. In a well-designed connector, some swage is necessary for the pin to fit inside the receptacle, as it creates a force which keeps the surfaces in contact with each other. Swipe describes the sliding action of the mating surfaces as they force their way past each other in the completion of the mating process. If all three of these processes occur during the mating cycle, the contact surfaces will be wiped clean of the thin oxide layer and a reliable connection will be established and maintained.

As is usually the case in engineering, there are tradeoffs to be made here. Separable connectors are characterized by the fact that they spend a significant portion of their lives in the unconnected state, which means that the mating surfaces of the connector are exposed to oxygen, dirt, humidity and moisture for extended periods of time, all of which tend to increase the challenge of establishing a reliable connection on the next mating cycle. One of the solutions is to increase the force between the two contacting surfaces, or the swaging force; this causes more surface oxide or other impurities to be scraped off during the mating process. However, this same force also causes increased contact wear, because the same abrasion which rubs off surface oxides also rubs off some of the good surface material, and decreases the number of mating cycles possible.

Another solution is to make the mating surfaces of precious metals, particularly gold. Gold does not oxidize very readily; in fact, it takes many years of humidity and oxygen to form a layer of gold oxide thick enough to cause a problem. The problem here is that gold is very soft and will rub off very easily, so the contact force must be decreased. Gold is also very expensive, so the manufacturing process must be precise enough to deposit the gold exactly where needed, and in the needed thicknesses, and no more.

In higher voltage connections (>48V), the general approach is simply to use metals which are very hard and have excellent spring characteristics. The classic example is a lamp cord or appliance plug and its corresponding receptacle. These contact surfaces do go through the contact, swage, swipe process, and a relatively reliable connection is formed. The connection formed actually has a significant amount of contact resistance, but in most cases this can be overcome by the voltages applied and by over-designing the contact surface area to account for this. However, some readers will be familiar with the common manifestation of this parasitic contact resistance, as it tends to heat up the connection ends of extension cords when a large amount of current is passed through the cord. It is another classic case of design tradeoffs, where the connection formed is adequate for most applications, but can be a problem in extreme cases.

In electronic connectors, where the contact resistance must be very small (<50 mΩ), the general solution is to use precious metals and reasonable contact forces. But again the tradeoffs: high contact forces reduce contact life, but low contact force means using precious metals, and large amounts of precious metals increase the price. High contact forces lead to another problem, which is often termed insertion force. If each pin of a 5-pin connector requires an insertion force of 50 grams, it only takes a force of 250 grams to make the connection. However, if each pin of a 100-pin connector requires a 50-gram insertion force, the connection requires a force of 5 kg, which may be very difficult to accomplish in awkward orientations or crowded enclosures, and which force will quickly crush any pins which are not properly aligned with their receptacle.

Connectors designed for rugged environments, such as military applications, generally use waterproof housings; when coupled, these connectors may be immersed without damage. However, the tradeoff is that such connectors are very expensive, and somewhat bulky and heavy. Other connectors for rugged environments, such as in the engine compartment of an automobile, are designed to be connected most of the time, and are only intended to go through a very limited number of mating cycles. They also provide two very desirable features: they are locked and they are keyed. The locking feature provides that they cannot accidentally separate due to vibration, but with the proper action, the lock can be defeated and the connector separated. The keying feature provides that the two connector halves cannot be mated in the wrong orientation. Also, each connector pair is different, so that the connectors will only be re-coupled with their corresponding other halves when the repair is complete.

5.2.3: Connector Materials

Housing Materials

The desirable characteristics for connector housing materials are the following:

1. low density, so that it will not add too much weight
2. high dielectric strength, to withstand the rated voltage
3. low dielectric constant, so that it will not add too much parasitic capacitance
4. high volume resistivity, so that it will be an excellent insulator
5. low moisture absorption, so that it will stay an excellent insulator
6. high arc resistance, so that arcs cannot easily form over its surface
7. high maximum service temperature, so it can be used over a wide temperature range
8. high tensile strength and high impact strength, so that it will not break easily
9. low flammability, so it will not readily combust under adverse conditions.

These requirements often go against each other, so the ideal material for a connector depends heavily on its application. Table 5.1 gives the relative rankings of several materials commonly used in connector housings.

Pin Materials

The desirable characteristics for pin materials are the following:

1. medium modulus of elasticity, so that it will be relatively stiff, yet somewhat springy
2. high yield strength, so that it will not break easily
3. high electrical conductivity
4. high thermal conductivity

Modulus of elasticity, yield strength and thermal conductivity are mechanical and thermal properties which are defined in the Glossary of this book. Table 5.2 gives some of these properties for several materials commonly for making connector pins.

The desirable characteristics of pin plating materials are the following:

1. high corrosion resistance
2. high electrical conductivity
3. high wear resistance
4. low cost

	Poly-amide (Nylon 6/6)	Polyphe-nylene Sulfide (PPS)	Epoxy (Novo-lac)	Phen-olic	Polycarbo-nate	Acryloni-trile-butadiene-Styrene (ABS)	Polyethylene Terepthalate (PET)
Density	1.14	1.35	1.92	1.74	1.20	1.04	1.34
Dielectric Strength (kV/mm)	23.6	10.0	15.3	15.0	15.0	9.5	16.5
Dielectric Constant	4.6	3.6	4.6	5.0	3.0	3.1	3.8
Volume Resistivity (Ω*cm)	$1x10^{13}$	$1x10^{16}$	$1x10^{13}$	$2x10^{11}$	$8.2x10^{16}$	$1x10^{16}$	$1x10^{15}$
Moisture Absorption	High	Very low	Very low	Very low	Very low	Low	Very low
Arc Resistance @ 23˚C (s)	130	34	187	180	120	67	123
Maximum Service Temperature (˚C)	130	200	204	230	115	130	150
Tensile Strength (MPa)	77.2	3.5	72.4	51.7	62.0	29.1	3.34
Impact Strength	Medium	Very low	Very low	Very high	Very high	High	Very low
Flammability Rating	Good	Excellent	Excellent	Excellent	Good	Poor	________

Table 5.1. Typical connector housing materials and their properties.

	Alloy 260 Brass	Alloy 272 Beryllium-Copper	Alloy 510 Phosphor-Bronze	Alloy 638	Alloy 725	Alloy 762 Nickel-Silver
Nominal Composition	Cu 70 Zn 30	Cu 98.1 Be 1.9	Cu 94.81 Sn 5.0	Cu 95 Al 2.8 Si 1.8 Co 0.4	Cu 88.2 Ni 9.5 Sn 2.3	Cu 59.25 Zn 28.75 Ni 12.0
Modulus of Elasticity (GPa)	112	130	112	117	135	127
Yield Strength, Annealed (MPa)	70-220	109	150	410-470	180	200
Electrical Conductivity (Siemens/cm)	163K	128K	87K	58K	64K	
Thermal Conductivity (W/m ˚C)	121	120	69	41	54	42

Table 5.2 Typical connector pin materials and their properties.

Table 5.3 rates several materials commonly used for plating connector pins.

	Ag	Au	Ni	Rh	Sn	Au over Ni	Rh over Ni
Corrosion Resistance	Fair	Superb	Excellent	Fair	Fair	Excellent	Fair
Electrical Conductivity	Superb	Excellent	Fair	Poor	Fair	Excellent	Poor
Wear Resistance	Fair	Poor	Very good	Excellent	Good	Good	Excellent
Cost	High	Very high	Low	High	Low	Medium	Medium

Table 5.3 Typical pin plating materials and their properties.

Wire Materials

The desirable characteristics for wire materials are the following:

1. low cost
2. high electrical conductivity
3. high ductility, so that the metal can easily be drawn into wire
4. low resistance temperature coefficient, so that the resistance of the wire does not increase significantly with temperature
5. high flex life, so that vibration, shock and flexing do not break it easily
6. low oxidation resistance, so that pins attached to the wire will not have to go through a layer of high resistance oxide.

The only two materials in general use for connector wire materials are aluminum and copper, with copper being much more common. The reasons for this are found in the rankings for the top contenders in each of these characteristics, as found in Table 5.4. From these rankings, it can be seen that copper is in first place, with aluminum next, and silver and gold far behind.

	Copper	Aluminum	Copper over Steel	Silver	Gold
Cost	Low	Low	Low	High	Very high
Electrical Conductivity	Very high	High	High	Very high	Very high
Ductility	High	High	High	Very high	Very high
Temperature Coefficient of Resistance (TCR) (%/°C)	.44	.48	.52	.60	.40
Flex life	Good	Good	Very good	Very good	Very good
Oxidation Resistance	Poor	Fair	Poor	Poor	Superb
Density	8.96	2.70	8.2	10.5	19.3

Table 5.4 Typical wire materials and their properties.

Wire Insulation Materials

The desirable characteristics of wire insulation materials are the following:

1. low density, so that it will not add a lot of weight
2. high abrasion resistance, so that the wire can be dragged, rolled over, pulled through, and otherwise abused without damage
3. high temperature capability
4. low dielectric constant
5. high dielectric strength
6. high volume resistivity.

Table 5.5 summarizes these characteristics for several types of materials used for wire insulation. The materials are given in alphabetical order, not in order of their use in industry. The three most common wire insulation materials are PE, PVC, and PTFE.

Characteristic:						
Material	**Density**	**Abrasion Resistance**	**Tempera-ture Range**	**Dielectric Constant**	**Dielectric Strength (kv/mm)**	**Volume Resistivity (Ω-cm)**
Polyamide (Nylon)	1.14	excellent	very good	7.0	15.2	4.5×10^{13}
Polychloroprene rubber (Neoprene)	1.2	excellent	good	9.0	32.0	1.0×10^{11}
Polyester Terepthalate (Mylar)	1.4	good	very good	3.1	102.4	6.0×10^{16}
Polyethylene (PE)	0.94	poor	poor	2.3	18.9	1.0×10^{16}
Polyimide (Kapton)	1.4	excellent	excellent	3.5	213	1.0×10^{18}
Polytetrafluoro-ethylene (PTFE, or Teflon)	2.2	poor	very good	2.1	18.9	1.0×10^{18}
Polyurethane	1.2	very good	good	7.2	19.0	2.0×10^{11}
Polyvinyl chloride (PVC)	1.4	good	poor	6.0	15.8	2.0×10^{14}
Polyvinyl fluoride (Kynar)	1.7	fair	good	7.7	50.0	2.0×10^{14}
Silicone rubber	1.2	poor	fair	3.3	27.0	2.0×10^{15}

Table 5.5 Typical wire insulation materials and their properties.

CHAPTER SUMMARY

Discrete transistors and diodes are not common in electronic products, but when they are the best option, careful selection of the transistor or diode package is very important. Some of the properties that must be considered include hermeticity, weight, size, and power dissipation capability.

Connectors and connections have been improved much over the past decades, but they are still a weak spot in electronics. Nearly 2/3 of all failures in electronic systems can be attributed to a failure in a connection or connector. Connectors (separable connections) are made up of a housing or header, the pins, the wire, and the wire insulation. Each performs a different function and is made of different materials. Options for these materials, and their performance in several attributes, are summarized in Tables 5.1 through 5.5. As is usually the case, no single material is best for all applications.

FOCUS QUESTIONS

1. Give eight of the most common specifications for a connector.
2. Give the three grades for connectors (and other electronic devices) and their respective temperature ranges.
3. High insertion force makes it difficult to insert a plug into its matching receptacle, yet reliable connections depend on a relatively high insertion force. What is the problem if the insertion force is decreased?
4. Gold is an excellent metal for the contact points of connectors because it essentially does not oxidize. However, it has its problems. What are two of the problems of using gold for the contact points?
5. Which connector housing material property is the most important to consider for an application which requires high ruggedness?
6. Which connector pin material property is the most important to consider for an application which requires high current?
7. Connector pins must conduct current with low opposition, and must not oxidize much over time. Which of the pin plating materials is best for this requirement?

FOR ADDITIONAL READING

Blackwell, Glenn R., "The Electronic Packaging Handbook", Chapter 8, CRC Press, Boca Raton, FL, 1999.

Clifford, Martin, "Electronic Connections: Home and Car Entertainment Systems",

Prentice-Hall, Englewood Cliffs, NJ, 1987.
Fink, Donald G. and Christiansen, Donald, "Electronics Engineer's Handbook , Section 7, McGraw-Hill, New York, NY, 1989.
Harper, Charles A., "Electronic Packaging and Interconnection Handbook", Chapters 3, 4, McGraw-Hill, New York, NY, 1997.
Matisoff, Bernard S., "Handbook of Electronics Packaging Design and Engineering", Chapter 13, Van Nostrand Reinhold, New York, NY, 1990.
Pecht, Michael, "Handbook of Electronic Package Design", Chapter 5, Marcel Dekker, New York, NY, 1991.
Seraphim, Donald P., R.C. Lasky, and C. Li, "Principles of Electronic Packaging", Chapter 7, McGraw-Hill, New York, NY, 1989.

CHAPTER 6: HISTORY, SUBSTRATES AND PATTERNING

CHAPTER OVERVIEW

This chapter covers the following concepts:

- IC manufacturing is an incredible marvel in the world of manufacturing
- The significance of the development of the IC can hardly be overstated
- The number of transistors on an IC has grown consistently at an exponential rate ("Moore's Law")
- The Czochralski method is the primary method used for creating IC wafers
- Photolithography is a very complex process that is used to pattern IC wafers
- Major advances are needed to enable photolithography to meet future expectations

6.1: INTRODUCTION

In the history of manufacturing, there has never been a period of sustained, major advances in a particular product like there has been in the manufacturing of transistors and integrated circuits (ICs). In this and the next two chapters, we will cover the processes and materials used in making IC chips. Subsequent chapters will cover the processes and materials used in making IC packages, hybrid ICs, printed wiring boards (PWBs), PWB assemblies, and connectors.

6.2: IC MANUFACTURING (LEVEL 0): CHIPS

The process by which integrated circuits are made is, by all measures, one of the most complicated and precise operations in all the world. The level 0 product alone, without any of the packaging steps of level 1, can take as many as 500 process steps to complete, with alignment repeatability requirements of 2 nm, or about 8 silicon atoms. An IC can take as long as six months to pass through all the necessary process steps. Each step pushes the very limits of what can reliably be done.

IC fabrication facilities are some of the most expensive facilities on earth, costing over ten thousand dollars per square foot for the required clean room areas alone. The equipment required is also very expensive, with the most complicated pieces costing several million dollars each. In the end, the total facility can easily cost over two billion dollars, and that is just to build it and fill it with equipment. The building must then be staffed by one of the best-educated work forces of all industries, and it must be rigorously maintained. The fact that all this process can even be done, much less for the relatively small price paid for the finished product, is truly a modern manufacturing marvel. Automation has always played a very large part in the success of making ICs, and this continues to be the case today.

Our ability to peer into the world of the atom came from a need to see the structures commonly built in ICs. The AFM (Atomic Force Microscope) and the STM (Scanning Tunneling Microscope), modern atomic-level imaging devices, were not invented by microscope or SEM (Scanning Electron Microscope) companies, but by a semiconductor company: IBM. No other material has become so thoroughly studied, experimented with, purified, and understood as has silicon, in spite of its relative youth as a manufacturing material. Statistical process control methods, now heavily used in general manufacturing, have been widely used in IC manufacturing since the early 1970s. The manufacturing of ICs brings together the fields of chemical, electrical, industrial, manufacturing, and mechanical engineering, along with physics, chemistry, mathematics, statistics, and other fields.

Integrated circuits are the enablers of countless other advancements in our modern society, including transportation, finance, management, mining, construction, materials, and most notably, information. In terms of dollars, almost one-fourth of all manufacturing is now electronic, more than any other segment of production in the world, including agriculture, mining, and transportation. In short, the IC is at the heart of the modern society we live in, and it is hardly a subject to be passed over lightly. We will spend a few pages learning about the history of this modern marvel of electronics, the integrated circuit.

6.3: HISTORY OF THE IC

As was mentioned in chapter 1, the integrated circuit was invented in 1959, by two people at two different companies, at essentially the same time. For several years prior to this, many companies had been perfecting the process of making transistors, from the very uncertain and unreliable first transistor to the dependable and inexpensive mass-produced versions which were then being used in nearly all new circuit designs. The process involved creating areas of p-type and n-type electrical characteristics in the semiconductor substrate. Both inventors realized that other devices besides transistors could be created by the same type of process, and both described a method by which more than one transistor, along with more than one resistor, plus the required interconnections, could be fabricated on the semiconductor substrate. Both companies, Fairchild Semiconductor and Texas Instruments, and both inventors, Robert Noyce and Jack Kilby, are to be credited with inaugurating the modern era of the IC.

Throughout the history of IC development, the most important device on the IC has always been the transistor. Therefore, in describing how much circuitry has been put on a single IC, industry has always specified the number of transistors on the IC, ignoring other devices such as resistors, diodes and capacitors. In the early 1960s, ICs became available that had up to ten transistors per chip; this level of integration became known as small scale integration, or SSI. In the mid-1960s, the process improved to allow up to 100 transistors per chip, which became known as medium scale integration, or MSI. In the late 1960s this increased to 1000 transistors per chip, a level of integration referred to, not surprisingly, as large scale integration, or LSI. Over the next several years, the number of transistors per chip continued to increase, and in the late 1970s chips with over 100,000 transistors were available, a level of integration known as very large scale integration, or VLSI. This continual increase in the number of transistors per IC has proceeded on a steady pace from the early 1960s even until now. Observing this trend, in the early 1970s Gordon Moore (later a founder of Intel) commented on the fact that the number of devices per IC seemed to be growing at the rate of doubling every 18 months, an observation which has since been termed "Moore's Law." It should have been called much more accurately, "Moore's Observation," for it is certainly not a law, but it is true that this level of improvement has continued for these past fifty–plus years, bringing us to a level of being able to put over 100 billion transistors on a single IC. Various terms have been coined to describe some of the later levels of integration, including ultra-large scale integration (ULSI), but the current level is often still referred to as VLSI, having run out of adequate new descriptions.

The number of different standard (off-the-shelf) ICs currently available is well over 300,000, with hundreds of thousands more being made for captive markets (thus not being publicly available). The array is mind-boggling, and new ICs are constantly being designed and fabricated. Plans exist for ICs to reach ever higher levels of integration, clear through the year 2025. Whether or not the current levels of advancement can be sustained is not entirely clear, but it is certain that dramatic improvements are yet to occur. More will be said on this topic later, after we have laid some of the groundwork necessary to understand the challenges of the future for ICs.

6.4: SUBSTRATES (WAFERS)

The first step in making an IC is to make the substrate on which the IC is fabricated. Most IC substrates are wafers that are made by the Czochralski method, which is the method described here. Other methods exist, and are relatively common in growing non-silicon substrates, but the Czochralski method is predominant for silicon.

In chapter 3, we learned about the three main states of solid matter. The state of most interest in growing silicon wafers is monocrystalline. In an ideal monocrystalline material, every atom is spaced equally from all neighboring atoms, in all three dimensions (no dislocations), and there are only atoms of the desired material (no impurities). Each transistor on an IC needs to have consistent characteristics; this requires that the crystals from which the transistor is fabricated be without dislocation or impurity. The Czochralski method has been refined and improved many times over the years, and now yields the most defect-free of all materials: monocrystalline silicon with purities and dislocations rated in the ppt (parts per trillion).

The Czochralski method is depicted schematically in Figure 6.1. The crucible is made of high purity quartz (SiO2). It is surrounded by heater elements, a graphite susceptor, insulation, and a water-cooling jacket. At the start of a growth cycle, the crucible is filled with chunks of metallurgical-grade silicon, already about 99 percent pure. Once filled, the entire growth chamber and lift tower is sealed and an inert gas (usually Argon) fills the chamber to prevent unwanted reaction of the silicon with the atmosphere. The remainder of the process is highly automated and very tightly computer-controlled.

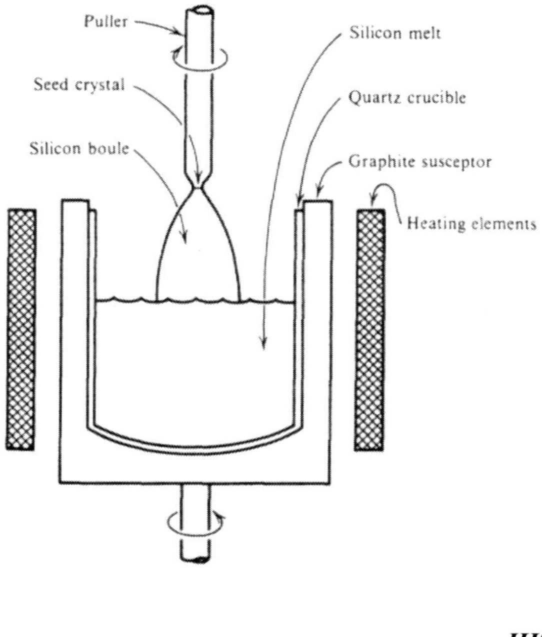

Figure 6.1 The Czochralski method for growing monocrystalline silicon. *From Ruska, W. Scot. 1987. Microelectronic Processing: An Introduction to the Manufacture of integrated Circuits. New York: McGraw-Hill. Courtesy of The McGraw-Hill Companies.*

The

crucible is heated until the silicon charge melts, after which it is cooled to just above the melting point of silicon (1420°C). A seed of the desired crystallographic orientation is then lowered to contact the top surface of the molten silicon. When the molten silicon, just barely above its melting point, comes into contact with the seed, it solidifies in the same orientation as the seed, and the crystal begins to grow. As it grows, the seed is lifted, and the solidifying crystal emerges from the molten silicon. The rate at which the crystal is pulled from the melt determines the diameter; 200 mm (8") and 300 mm (12") are the sizes still common today, but several companies are experimenting with still larger silicon wafers. Rates of 50 mm/hour are common in Czochralski crystal growth. Thus it would take approximately 12 hours to grow a crystal 600 mm (24") in length. The finished crystal is usually termed a *boule*.

The impurities in the molten silicon do not readily mix with the interface between the solid crystal and the molten material, and remain in the molten silicon until near the end. The final several centimeters of the boule contain the majority of the impurities and dislocations, and this part is not used for making wafers.

When the boule is harvested, it contains many irregularities in its surface, due to the growth process. The boule is ground on a lathe to give it a constant diameter, after which it is ready to be sliced into wafers. The wafer slicing process is actually a grinding process, using an inside-diameter saw (see Figure 6.2). It is desirable to keep the saw blade as thin as possible, as all the material removed in the wafer-slicing process is a loss. This saw is actually a thin band of diamond-coated metal, stretched tight to keep it flat. The wafers are cut to thicknesses of about 12 to 18 mils (.30 to .45 mm), which is the minimum necessary to give the wafers enough mechanical strength to withstand all the handling of the IC manufacturing process.

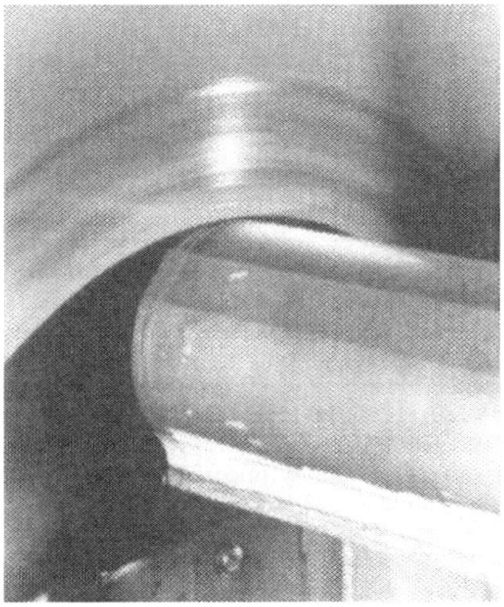

Figure 6.2 An inside diameter, diamond coated saw slicing a wafer from a boule. *From Harper, C. and R. Sampson. 1992. Electronic Materials and processes Handbook. New York: McGraw-Hill. Courtesy of The McGraw-Hill Companies.*

After the wafers are sliced, they are then ground and polished (see Figure 6.3) to give a very high mirror finish to the surface. Surface roughnesses of no more than about 100 nm are required to make ICs, for reasons we will soon come to understand. The finished product is a single-crystal wafer with an extremely flat surface, ready for the IC fabrication process.

The functions of IC manufacturing can be classified into three main categories.

1. Patterning (see next section)
2. Adding materials (see Chapter 7)
3. Removing materials (see Chapter 8)

6.5: PATTERNING

Patterning of ICs is done primarily by a process known as photolithography. The word means drawings (graphy) on stone (litho) using light (photo). This process is fundamental to the manufacture of ICs, and should be well understood by the reader. It is the process by which various regions are defined in the silicon; the various regions make up the parts of the transistors, resistors, and connections on an IC, as depicted in Figure 6.4.

6.5.1: The Process of Photolithography

Photolithography is basically a seven step process; six of these steps proceed as shown in Figure 6.5.

Figure 6.3 A two-sided, multi-wafer alpper and polisher. *From Harper, C. and R. Sampson. 1992. Electronic Materials and processes Handbook. New York: McGraw-Hill. Courtesy of The McGraw-Hill Companies.*

1. A thin layer of SiO_2 is grown on the surface of the wafer, using heat and either water vapor (steam) or oxygen in a temperature-controlled oven. The steam or oxygen reacts with the surface silicon to form this reaction product.
2. A layer of a liquid photosensitive material (photoresist) is deposited on the surface of the SiO_2. This photosensitive layer is then cured (hardened) at an elevated temperature.
3. The wafer/ SiO_2 /photoresist is exposed to UV light through a mask or "reticle". In a reticle, light passes through those areas where it is transparent, or else the light is blocked by those areas where the non-transparent portions of the reticle block the light. In positive photoresists, the portions of the photoresist that are exposed to the UV light soften. The portions of the photoresist which have not been exposed to the UV light remain hardened (cured).
4. The exposed photoresist layer is developed by immersion in the developing chemical; once immersed in the developing chemical, the softened (exposed) photoresist dissolves and is thus removed. This leaves openings in the photoresist where the reticle was transparent.
5. The SiO_2 layer is etched away where it has been exposed by the removal of the photoresist layer above it. The part of the SiO_2 still covered with photoresist resists the etching process.
6. The layer of cured photoresist is removed in a stripping operation.

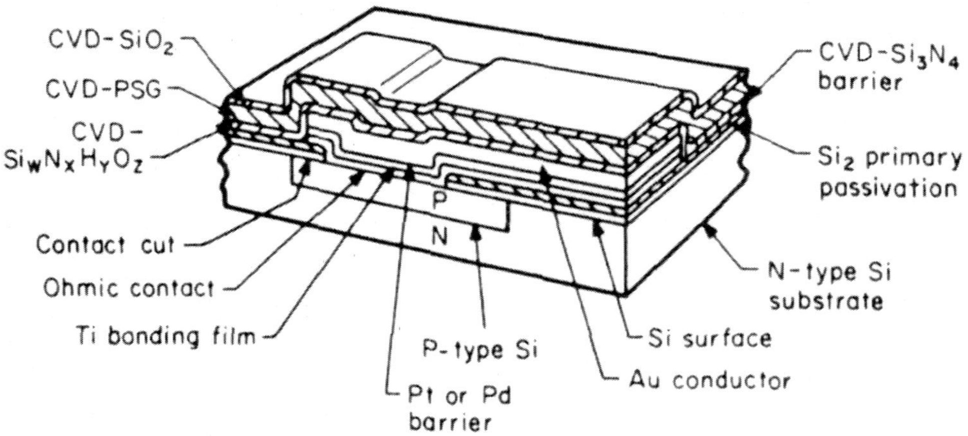

Figure 6.4 Various regions on an integrated circuit diode. *From Elliot, David. 1989. Integrated Circuit Fabrication Technology. New York: McGraw-Hill. Courtesy of the McGraw-Hill Companies.*

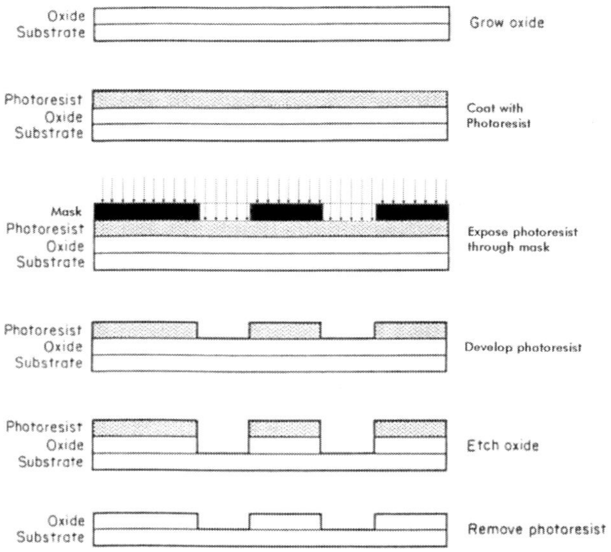

Figure 6.5 The six main steps of photolithography. *From Elliot, David. 1989. Integrated Circuit Fabrication Technology. New York: McGraw-Hill. Courtesy of the McGraw-Hill Companies.*

At this point, the desired manufacturing process (adding or removing material) is performed on the exposed silicon areas, while the remainder of the silicon is covered by the layer of SiO_2. There are many manufacturing processes that can be performed through this window; these will be discussed in the following sections. These manufacturing processes of adding or removing material are not part of the process of photolithography.

7. After the desired manufacturing process is complete, the layer of SiO_2 is chemically removed.

Each mask layer requires that this basic process of photolithography be repeated for that mask layer. Because ICs can have up to twenty-five mask layers, this process of photolithography is repeated many times.

The main advantages of photolithography are the following:

1. Many chips can be processed at the same time.
2. The resolution of the process is limited primarily by the wavelength of the light used, and this can be quite small.
3. Photolithography can be highly automated, reducing cost and improving quality.

6.5.2: The Tools and Materials of Photolithography

The main tool for photolithography is usually referred to as a stepper. This is because the tool uses a partial-wafer mask, or "reticle," to expose the photoresist, and this reticle must be stepped to sequential locations on the wafer to image the entire wafer. Figure 6.6 shows a schematic diagram of the exposure process, or the actual process of "stepping."

Modern photolithography tools comprise a mini-environment in which steps one to four are accomplished under completely automated control; transfer between these four process steps is performed by a robotic arm. Figure 6.7 is an example of one of these tools. Such a tool is very complex, and costs up to 100 million dollars.

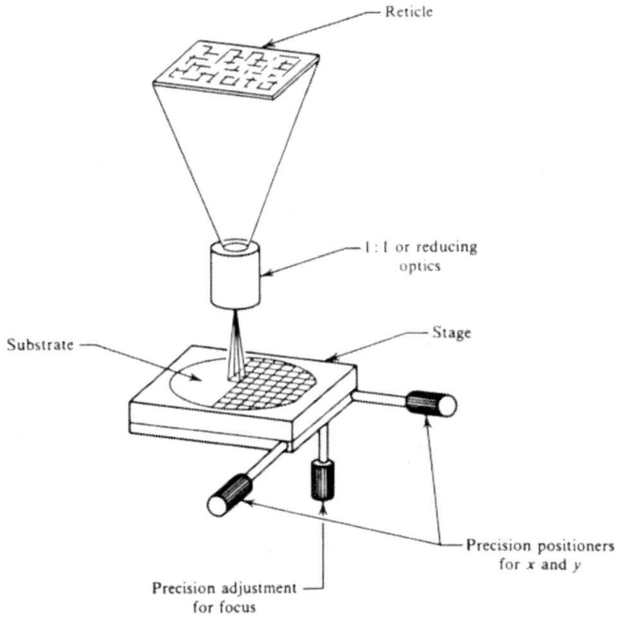

Figure 6.6 The stepping process in photolithography. *From Ruska, W. Scot. 1987. Microelectronic Processing: An Introduction to the Manufacture of integrated Circuits. New York: McGraw-Hill. Courtesy of The McGraw-Hill Companies.*

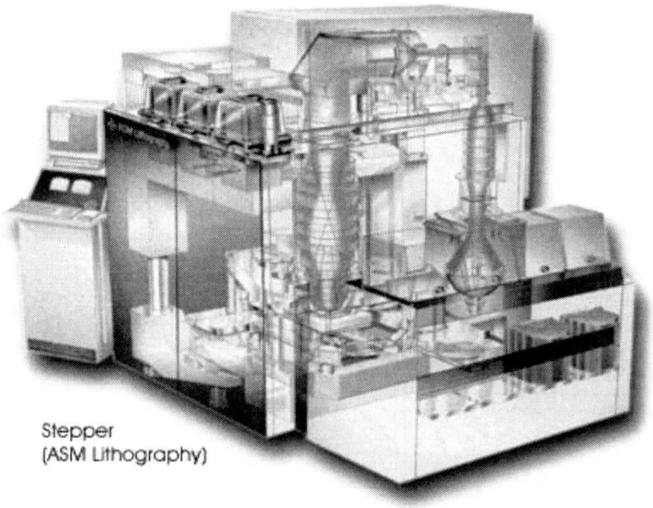

Figure 6.7 A photolithographic stepper. *Courtesy of ASM Lithography, Tempe, Arizona.*

The photoresist used is a hydrocarbon-based photopolymer which is a liquid when dispensed onto the wafer. This is usually done in a spinning operation, in which the wafer is rotated or spun while the photopolymer is dispensed onto the wafer (see Chapter 7). The photopolymer is then semi-hardened by an elevated temperature bake; it is then ready for exposure in the stepper.

The room in which the photolithography is performed must be fully temperature and humidity controlled. Any variations in these parameters can cause the alignment to go off, or the photoresist steps to have unwanted variations.

6.5.3: Issues in Photolithography: λ of Process, λ of Light Sources, Optics, Cost

The smallest feature possible in a photolithography process is termed the resolution of the process, and is usually designated with the symbol λ. Although resolution is directly dependent on the wavelength of light used in the photolithographic process, the λ of the process is not the same as the wavelength of the light, (which is also designated λ). This dual meaning of the symbol λ is unfortunate, but well ingrained in the IC industry. It is helpful if we think of them as λ of the process (resolution), and λ of the light (wavelength).

As was reviewed in section 6.3 on the history of ICs, the past several decades have seen continual reduction in the size of transistors on ICs. The technology of photolithography is one of the main areas that must be significantly improved with each generation of smaller electronic devices. Each generation is usually about $\sqrt{2}$ smaller in both the x and y dimensions, yielding a doubling of the number of devices in the same unit area. Thus, every two generations requires that the photolithography equipment be capable of resolution two times better than the previous generation. One of the biggest challenges in this continual effort to shrink geometries is the wavelength of the light used to do the photolithography. There is a fundamental physical limit that the resolution cannot be better (smaller) than ½λ (λ = wavelength of light source). Present technology allows us to approach this fundamental limit and achieve a resolution approximately equal to ½λ, so a process with a minimum line width, or process resolution, of 25 nm, would require a light source of 50 nm. Visible light is from about 400 to 700 nm, so 50 nm is in the extreme ultraviolet (EUV) region. But there are other requirements on the light source, besides the wavelength.

As the light passes through the reticle, it is desirable that it not spread. This requires that the light be collimated and monochromatic; this is represented by the parallel arrow-tipped lines in the expose step of Figure 6.5. Collimated light has all its photons traveling in parallel and in phase with each other, so the photoresist sees the exact same pattern as the reticle, without spreading or distortion of the lines. Monochromatic light is of a single (mono) color (chromo) or wavelength. Such light is difficult to come by, and is only truly available from lasers. However, other sources have been pressed into service successfully, and until the .35 µm generation, most of the UV light necessary was created using mercury vapor lamps and advanced optics. The 25 nm generation typically uses excimer lasers as the source, but this light source is much more expensive.

The wavelength emitted by a light source depends primarily on the material from which the light source is made. Thus, we cannot simply "tune" most light sources to emit the frequency we need, any more than we can change the molecular or atomic structure of the material from which it is made. The wavelengths of light available to us are constrained by what sources we have been able to discover and develop into practical sources. Presently, there are not many sources of light below 200 nm, and even fewer are practical.

Research has been conducted by several companies, for several years, on electron beam (e-beam) and x-ray lithography. Some very significant advances have been made, and record-setting dimensions below 15 nm have been achieved in the lab, but as of this writing, no production devices have yet been developed at this resolution.

6.5.4: Dimensions on ICs

It would probably be wise at this point to interject a small primer to familiarize the reader with the dimensions used in most literature on ICs. Unfortunately, it is a hodgepodge of both English and metric units, with overlaps and long-standing tradition in the use of each. Table 6.1 summarizes these units, but a bit of explanation is in order.

Unit	Symbol	Dimensions Measured	inch	mil	mm	µm	nm	Å
mil	mil	chip length, width, area	.001	1	.0254	25.4	25.4×10^3	254×10^3
millimeter	mm	chip length, width, area	.0394	39.4	1	1000	1.0×10^6	10×10^6
micrometer or micron	µm	device features, process resolution	39.4×10^{-3}	.0394	.001	1	1000	10,000
nanometer	nm	device features, light wavelengths	39.4×10^{-6}	39.4×10^{-3}	1.0×10^{-6}	.001	1	10
Angstrom	Å	thicknesses, light wavelengths	3.94×10^{-6}	3.94×10^{-3}	0.1×10^{-6}	.0001	.1	1

Table 6.1 Common dimensions used in IC fabrication.

Some helpful hints for remembering the dimensions of Table 6.1: human hair is about 2 to 3 mils in diameter, or about 50 to 75 μm. The radius of a silicon atom is about 1.46 Å; spacing between silicon atoms in a crystalline structure is about 4 Å. There are about 25 μm in a mil, and about 40 mils in a mm.

Long-standing tradition is often evidenced when discussing process resolution and wavelengths of light used. Physics has been around much longer than photolithography, and their established units for the wavelength of light are the nm or Å. When ICs were first made, the process resolution was in the range of 10 μm, so it made sense to use microns to measure it. Recently, however, the process resolution has shrunk dramatically, and nm are now commonly used.

6.5.5: The Future of Present Photolithographic Methods

One of the two main limitations to present methods of photolithography is the registration required for the necessary resolution. Registration requirements are typically one quarter the resolution of the process; thus a process with a resolution of 25 nm requires a registration of 6.25 nm. Such incredibly small registration requirements are increasingly difficult (read: expensive) to achieve.

The other main limitation is the wavelength of the light available for photolithography. Many companies have been working for several years on a solution to this problem, and at the present, EUV steppers are available, but are extremely expensive. These steppers are capable of resolution of about 20 nm.

Other fundamental limits exist which are beyond the scope of this discussion. Where we are with respect to them is the subject of much debate, as is the discussion as to whether they are true fundamental limits, or merely more perceived limits. It is very likely that the past history of continued shrinking will be slowed eventually not by fundamental limits, but by practical limits. That is, eventually the necessary equipment and facilities will cost so much to buy and to build that the ICs themselves will become unreasonably expensive. As we near this point, Moore's Law will inevitably no longer hold, and the rate of increase in the number of transistors per unit area will begin to slow. How far away is this? Current projections are mostly conjecture, because so many things can happen between now and then, but the general consensus seems to be that this slowing will occur sometime after the year 2025, and probably before the year 2030.

CHAPTER SUMMARY

There is no industry in all of history that has so consistently improved over such a short time as the electronics industry. Each succeeding generation has been smaller, faster, lighter, less expensive, and has used less power. This improvement has been primarily due to the improvement in ICs, which for over 5 decades have followed the observation known as "Moore's Law". At the heart of these incredible advances have been the improvements made to the process of photolithography.

The predominant material used for making ICs is silicon, which is grown in crystals via the Czochralski method. These large crystals are sliced into wafers, which are then polished.

Patterning of each layer of an IC is done by the complex and expensive process known as photolithography. At the heart of this process is the "stepper", a very expensive and very precise tool used heavily throughout the industry. Understanding the process of photolithography is essential to understanding how ICs are made.

FOCUS QUESTIONS

1. Why were today's most powerful microscopes (the STM and the AFM) invented by IC companies?
2. IC manufacturing facilities are some of the most expensive and demanding facilities in the world. How are we able to build them and make them profitable?
3. What is a boule?
4. What does VLSI (very large scale integration) mean?
5. Describe in your own words the Czochralski method of growing ultrapure silicon crystals.
6. We now know more about silicon then we do about any other material on earth. Why?
7. Describe in your own words the basics of the photolithography process.
8. What is a "stepper"?
9. Several years ago, photolithography was done by exposing the entire wafer all at once, through a full-wafer mask. Modern photolithography uses steppers, which are much more expensive, and which only expose a portion of the wafer at a time through a reticle (so they are slower). Why did we change to something more expensive and slower?
10. What is the relationship between the λ of a photolithography process and the λ (wavelength) of light?
11. What are three of the requirements of a good light source for photolithography?
12. A common sheet of paper is about 2 mils thick. What is the thickness in μm?
13. A common wavelength used in making ICs is 320 nm. What is the wavelength in μm?
14. A common gate oxide thickness is 20 nm. What is the thickness in Å?

15. What is one of the major obstacles to using shorter wavelength light sources for photolithography?

FOR ADDITIONAL READING

Elliott, David J., "Integrated Circuit Fabrication Technology", Chapter 1, McGraw-Hill, New York, NY 1989.

Ghandi, Sorab K., "VLSI Fabrication Principles: Silicon and Gallium Arsenide", John Wiley & Sons, New York, NY, 1994.

Harper, Charles A., and R.M. Sampson, "Electronic Materials & Processes Handbook", McGraw-Hill, New York, NY, 1994.

Pecht, Michael, R. Agarwal, P. McCluskey, T. Dishongh, S. Javadpour and R. Mahajan, "Electronic Packaging: Materials and Their Properties", CRC Press, Boca Raton, FL, 1998.

Querk, Michael, and J. Serda, "Semiconductor Manufacturing Technology", Prentice-Hall, Upper Saddle River, NJ, 2001.

Ruska, W. Scot, "Microelectronic Processing: An Introduction to the Manufacture of Integrated Circuits", McGraw-Hill, New York, NY 1987.

CHAPTER 7: ADDING MATERIALS

CHAPTER OVERVIEW

This chapter will cover the processes used to add materials to all or to selected portions of the wafer. These processes include:

- Dispensing and spinning
- Oxide growth
- Sputtering
- Evaporation
- Chemical vapor deposition (CVD)
- Diffusion
- Ion Implantation
- Molecular beam epitaxy (MBE)

7.1: DISPENSING AND SPINNING

Starting at the lowest-tech method of adding material, we consider first the addition of the photoresist. As mentioned before, this liquid is dispensed onto the wafer as the wafer is spun; the spinning provides that the film thus formed is uniform in thickness, except along the edges of the wafer (see Figure 7.1).

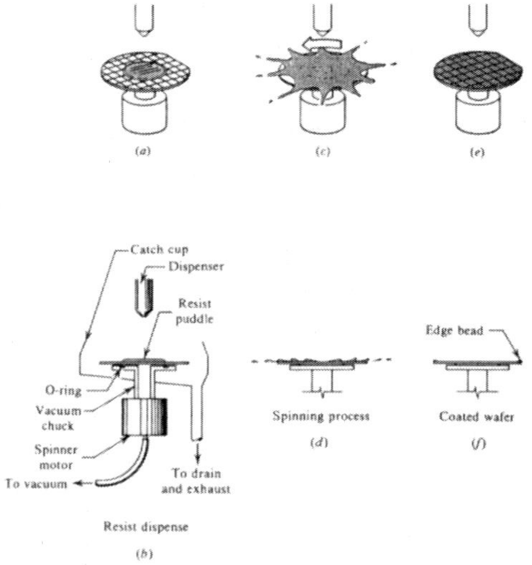

Figure 7.1 Spin coating of photoresist on the wafer. From Ruska, W. Scot. 1987. Microelectronic Processing: An Introduction to the Manufacture of Integrated Circuits. New York: McGraw-Hill. Courtesy of the McGraw-Hill Companies.

Photoresist is not usually added selectively. After spinning, the wafer is baked to harden the photoresist.

7.2: OXIDE GROWTH

The growth of oxide on the surface of the wafer was briefly mentioned in chapter 6, which discussed the process of photolithography. Growing oxide on the wafer is one of the most basic of all IC fabrication processes. Oxide serves both as an insulator between layers of wiring, and as a resist material in the process of photolithography. Oxide is grown in an oxygen or steam atmosphere, at temperatures between 950 and 1250°C. The silicon of the wafer reacts with the oxygen in the atmosphere to form a very stable oxide. Thicknesses vary from as little as 100 Å to as much as 10,000 Å, depending on the need. This process is usually performed in large diffusion ovens which can handle as many as 200 wafers at a time. Figure 7.2 gives a schematic of one of these ovens.

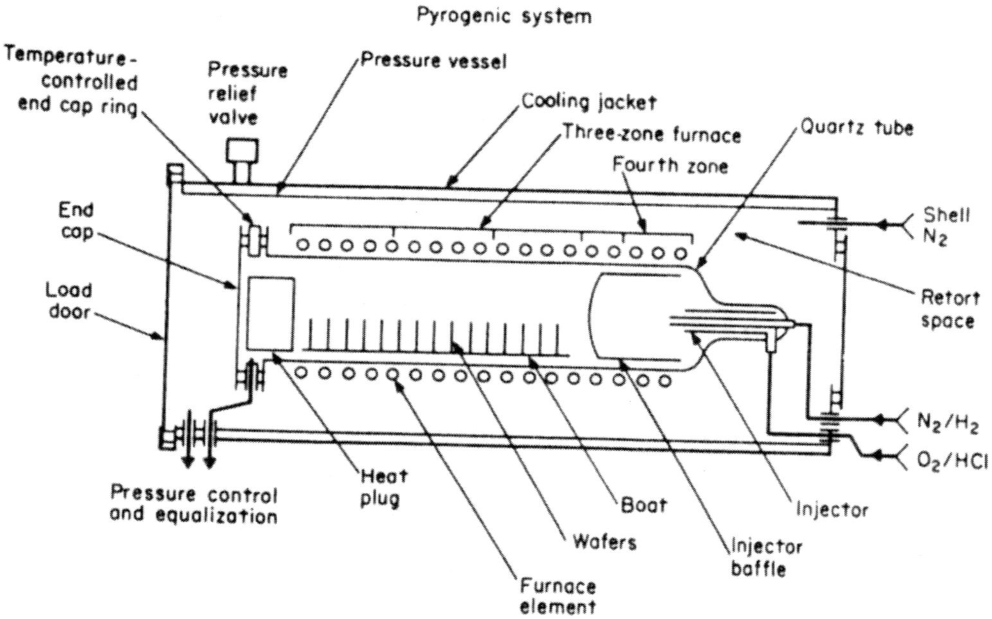

Figure 7.2 A high-pressure oxidation furnace. *From Elliot, David. 1989. Integrated Circuit Fabrication Technology. New York: McGraw-Hill. Courtesy of The McGraw-Hill Companies.*

7.3: SPUTTERING

Another material added to the entire wafer is the metal used for vias, or layers of wiring. These metals are typically aluminum and copper for wiring layers, and tungsten for the vias, plus other metals for specialized applications. These metals are added in a sputtering or an evaporation process. Sputtering is depicted in Figure 7.3, and is basically accomplished by the following steps:

1. ionizing an inert gas, such as argon, by exciting it with an electric field
2. shooting this beam of ions toward a target of the desired metal
3. dislodging atoms of the target with this ion beam
4. condensing these target atoms onto the wafer surface.

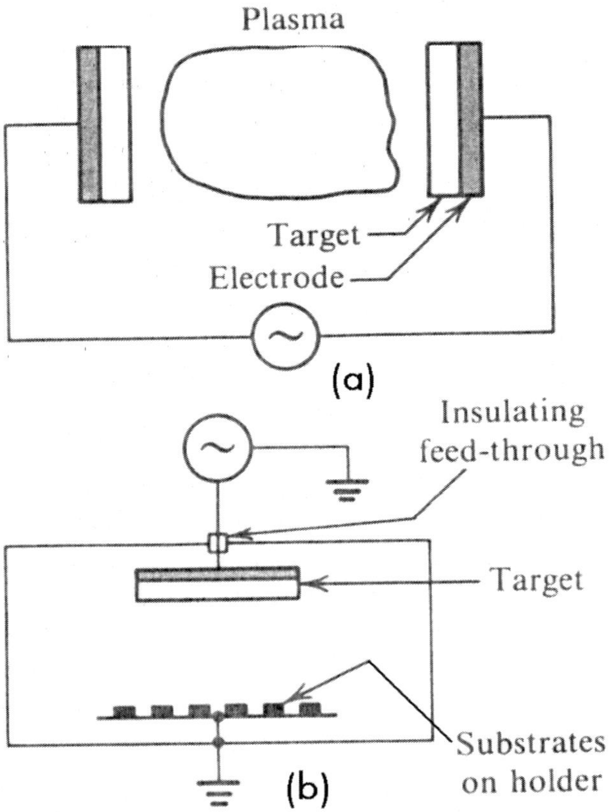

Figure 7.3 The sputtering process. (a) A symmetrical ac discharge, (b) A practical sputtering configuration including a target electrode and a substrate holder. *From Ruska, W. Scot. 1987. Microelectronic Processing: An Introduction to the Manufacture of Integrated Circuits. New York: McGraw-Hill. Courtesy of the McGraw-Hill Companies.*

This is done in a high vacuum to avoid the presence of impurities and other undesired atoms. Metals commonly added by sputtering include aluminum, copper, metal silicides, gold, titanium, tungsten, and molybdenum.

7.4: EVAPORATION

Evaporation is depicted in Figure 7.4, and is accomplished by enclosing the wafer and desired metal in a very high vacuum and heating the desired metal. This causes the metal to sublimate, or evaporate; these evaporated atoms of the desired metal are then drawn onto the wafer. Aluminum and other metals with relatively low melting points are the materials commonly added by evaporation.

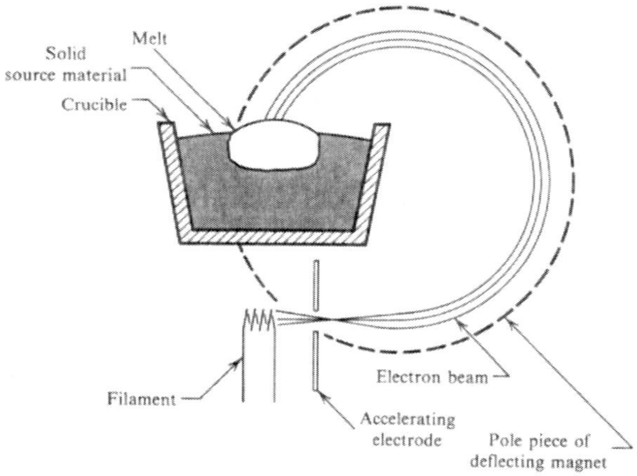

Figure 7.4 The evaporation process using electron-beam heating. *From Ruska, W. Scot. 1987. Microelectronic Processing: An Introduction to the Manufacture of Integrated Circuits. New York: McGraw-Hill. Courtesy of the McGraw-Hill Companies.*

The result of the processes of sputtering or evaporation is a thin layer of metal across the entire wafer, including in the trenches formed for some processes. It is also important in the process of depositing these metals that the vertical steps be adequately coated as well (see Figure 7.5), or the connections will be incomplete.

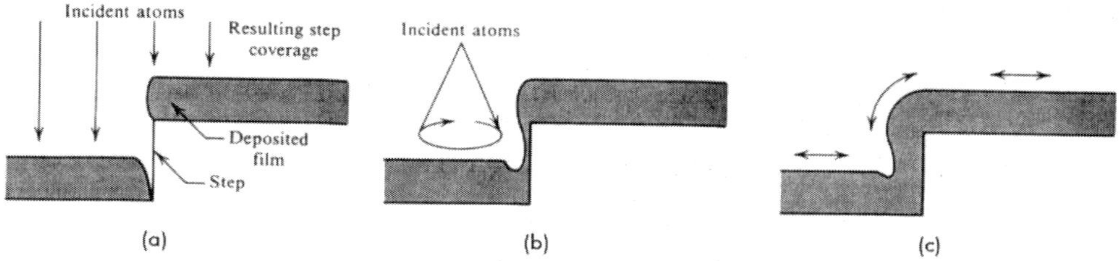

Figure 7.5 Examples of step coverage in the deposition process. (a) Incident atoms arriving only perpendicular to the substrate result in no step coverage, (b) Some stirring in the direction of the arriving atoms provides some step coverage, (c) Additional stirring results in adequate step coverage. *From Ruska, W. Scot. 1987. Microelectronic Processing: An Introduction to the Manufacture of Integrated Circuits. New York: McGraw-Hill. Courtesy of the McGraw-Hill Companies.*

7.5: CHEMICAL VAPOR DEPOSITION (CVD)

Chemical Vapor Deposition (CVD) involves the use of chemical gases that are combined and separated by heat into other molecules. The desired molecules are deposited on the surface of the wafer, either over the entire wafer, or in the areas previously patterned through photolithography. Some of the materials that are deposited in this manner include SiO_2, silicon nitride (Si_3N_4), polycrystalline silicon (poly), and aluminum oxide (Al_2O_3). CVD is also used to form films of many other materials including other insulators (tantalum oxide, titanium oxide), semiconductors (type IV, type III-V, and type II-VI), conductors (the same as listed above under Evaporation), and superconductors (usually compounds of niobium). There are several feeder gases that are used for the required chemical reactions, including silane, dichlorosilane, phosphine, diborane, and arsine, all of which are very toxic, flammable and corrosive, and must be handled with extreme care. It should be mentioned here that CVD is also used in the optical fiber business to make the glass rods from which the fiber is pulled. Several types of CVD reactors are depicted in Figure 7.6.

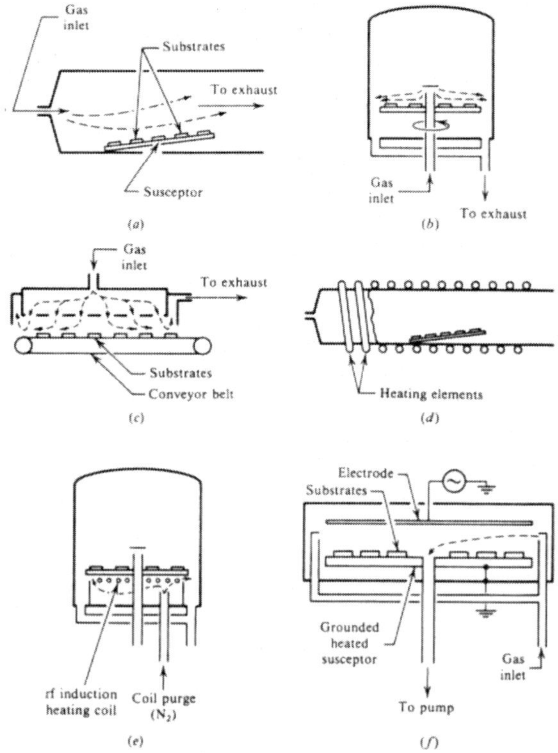

Figure 7.6 Types of CVD reactors. (a) Horizontal reactor, (b) Vertical reactor, (c) Continuous flow reactor, (d) Hot-walled reactor (horizontal), (e) Cold-wall reactor (vertical), (f) Plasma-entranced reactor. *From Ruska, W. Scot. 1987. Microelectronic Processing: An Introduction to the Manufacture of Integrated Circuits. New York: McGraw-Hill. Courtesy of the McGraw-Hill Companies.*

7.6: PHOTOCHEMICAL VAPOR DEPOSITION (PCVD)

The process of photochemical vapor deposition proceeds much as it does in CVD, described in the previous section, except that the feeder chemicals are made to react through the introduction of light, usually UV. This provides for low reaction temperatures and outstanding step coverage, as shown previously in Figure 7.5. The characteristics of low reaction temperature and excellent step coverage are desirable for many process steps.

7.7: DIFFUSION

Diffusion is usually performed on patterned areas of the silicon wafer, specifically for doping those areas we wish to make p- or n-type silicon. Diffusion dopants for p-type silicon are usually aluminum, boron or gallium (Column III); dopants for n-type silicon are usually antimony, arsenic, or phosphorous (Column V). Diffusion is accomplished by heating the wafer in an enclosed oven, then introducing the desired dopant into the oven at a known concentration for a fixed period of time. With tight controls over concentration, temperature and time, the dopant diffuses into the patterned areas of the wafer a fixed amount, creating the desired p- or n-type areas. Ovens used for diffusion are similar in most ways to those used for oxidation growth.

Diffusion is an isotropic process, meaning that the diffusion occurs at the same rate in all directions. Figure 7.7 shows what this means in making doped areas; if the desired area is to be 1 μm square and 1,000 Å deep, the opening through which the diffusion occurs must be made smaller by .2 μm in order to keep the diffusion within the desired 1 μm area, since it will diffuse laterally 1000 Å (.1 μm) in each direction as it diffuses 1000 Å downward.

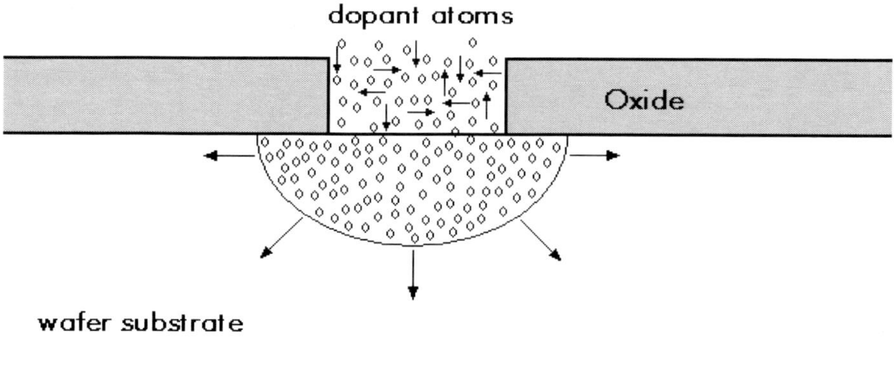

Figure 7.7 Diffusion progresses in all directions equally.

It also does not provide sharp edges in the profile of the diffusion area. In spite of this undesired isotropic nature, diffusion is used where feasible, due to its low cost. However, lately diffusion has generally become impractical where tight geometries are required, due to the very small dimensions necessary, and the difficulty in adapting an isotropic process to meet these demands.

7.8: ION IMPLANTATION

Ion implantation is the process that has basically replaced diffusion in applications of tight geometries. Its main advantage is that it is an anisotropic process, meaning that it proceeds primarily along a single axis. As shown in Figure 7.8, this provides a processed area which is the same in the x dimension as the opening, with straight walls in the profile. Although it is much more expensive than diffusion, it dominates primarily for its anisotropy.

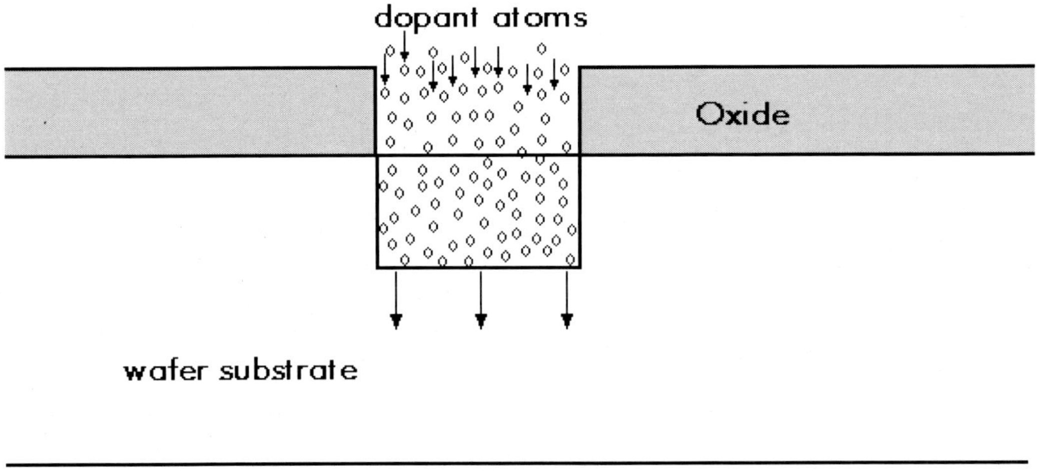

Figure 7.8 Ion implantation progresses primarily in a downward direction.

Ion implantation is accomplished by using an ion source of the desired atom, accelerating these ions toward the patterned silicon wafer, and driving the ions into the wafer. The depth to which the ions are driven depends on their energy, or voltage; the rate at which the ions are implanted depends on the current. Elements typically implanted include arsenic, boron, antimony and phosphorous. Figure 7.9 shows a schematic of an ion implanter.

Ions of the desired material impart their energy to the substrate, causing the substrate atoms to be displaced and creating a local area that is amorphous instead of monocrystalline. Because this is highly undesirable, the damage caused by ion implantation must be repaired by a subsequent process, known as annealing. The process of annealing consists primarily of raising the temperature of the substrate sufficiently (but not to the melting point) to allow the atoms to relocate into the monocrystalline state (a lower state of energy), after which the wafer is slowly cooled.

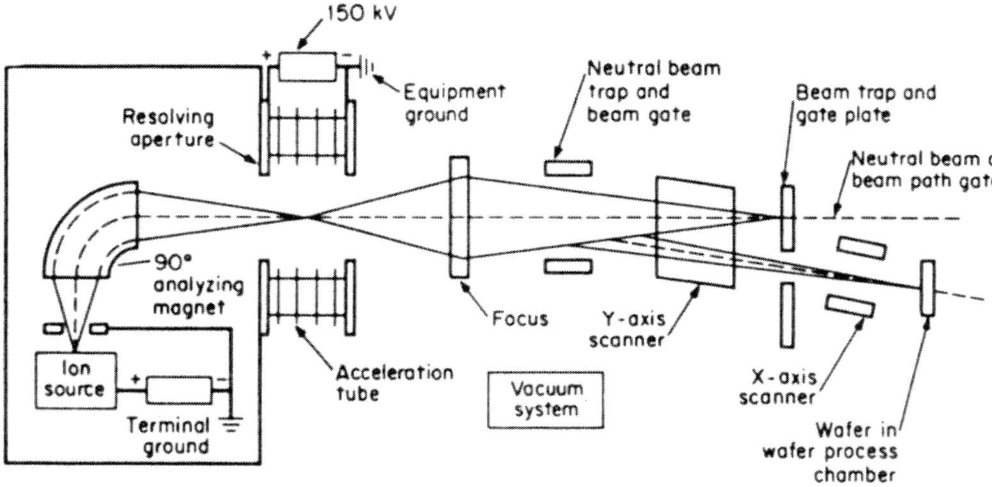

Figure 7.9 An ion implanter. *From Elliot, David. 1989. Integrated Circuit Fabrication Technology. New York: McGraw-Hill. Courtesy of The McGraw-Hill Companies.*

7.9: MOLECULAR BEAM EPITAXY (MBE)

This process uses a beam of the desired molecules to form a very thin layer on the surface of the substrate. The word epitaxy is used to describe where this layer forms; *epi* refers to the upper or outermost layer, as in *epi*dermis; *taxy* refers to the arrangement or order of the layer, as in *tax*onomy. The process takes place in an ultra-high vacuum, where an open, heated crucible of the desired material is made to evaporate the desired material. The beam of material thus created is drawn to the substrate thermally, and is controlled by an aperture placed between the crucible and the substrate. This process is primarily used to grow extremely thin layers of materials with very tightly controlled dopant profiles. Figure 7.10 gives a schematic of this process.

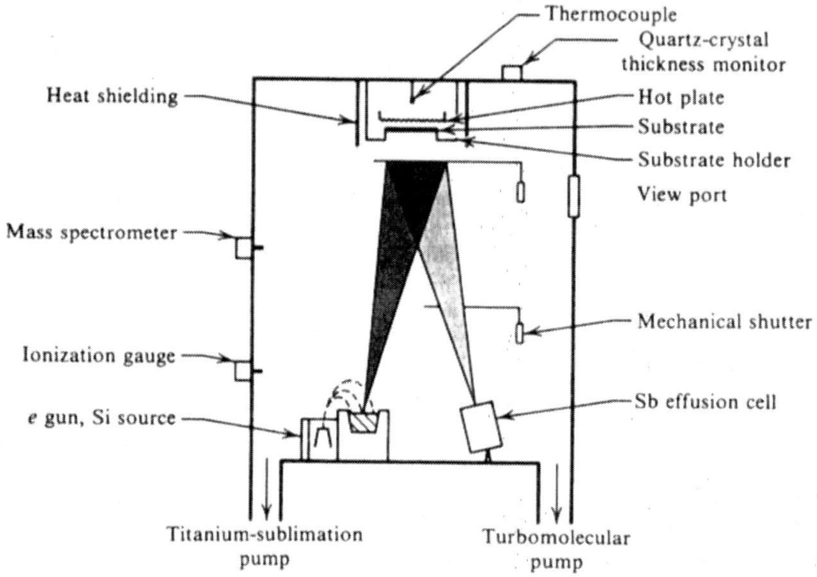

Figure 7.10 A molecular beam epitaxy apparatus. *From Ruska, W. Scot. 1987. Microelectronic Processing: An Introduction to the Manufacture of Integrated Circuits. New York: McGraw-Hill. Courtesy of the McGraw-Hill Companies.*

CHAPTER SUMMARY

The process of making ICs on a wafer includes both adding and removing materials. This chapter has focused on eight of the processes most commonly used to add materials to the wafers. These processes include:

1. Dispensing and spinning, which is most commonly used to add photoresist. This is a low-tech process which is robust and highly effective.
2. Oxide (SiO2) growth, which is usually done in an oven essentially the same as a diffusion oven, and allows many wafers to be processed at a time.
3. Sputtering, which is usually used to add different metals to the entire wafer. This process requires a vacuum.
4. Evaporation, which is also used primarily to add different metals. This process also requires a vacuum.
5. Chemical vapor deposition (CVD), which can be used to add almost any desired material, provided that the appropriate feeder chemical can be obtained. This is another process which requires a vacuum.
6. Diffusion, which can be done in a large furnace holding many wafers at the same time. This process is very old, very well understood, and relatively inexpensive, but is isotropic.
7. Ion implantation, which requires a very expensive and large piece of equipment, can only do a few wafers at a time, and requires a very high

vacuum. However, it is anisotropic.

8. Molecular beam epitaxy (MBE), which can be used to add several kinds of material to a wafer. This is another process which requires a vacuum.

FOCUS QUESTIONS

1. What is one of the major advantages of the dispensing and spinning method of adding materials?
2. What are the two main uses of silicon oxide in making IC's?
3. What kinds of materials are commonly sputtered onto a wafer?
4. Describe the problem that results from poor step coverage.
5. Many types of feeder gases are used with the CVD process. What are some of the problems associated with these gases?
6. Diffusion is much less expensive than ion implantation, yet is rarely used any more. Why?
7. Describe why annealing is often necessary after ion implantation.
8. The MBE process requires a high vacuum to be successful. Why is this so? What other methods of adding materials also require a vacuum?
9. What is the advantage of an anisotropic process, compared to an isotropic process?
10. Doping a wafer is done either by diffusion or by ion implantation. Which is an isotropic process and which is an anisotropic process?

FOR ADDITIONAL READING

Elliott, David J., "Integrated Circuit Fabrication Technology", Chapter 1, McGraw-Hill, New York, NY 1989.

Ghandi, Sorab K., "VLSI Fabrication Principles: Silicon and Gallium Arsenide", John Wiley & Sons, New York, NY, 1994.

Harper, Charles A., and R.M. Sampson, "Electronic Materials & Processes Handbook", McGraw-Hill, New York, NY, 1994.

Pecht, Michael, R. Agarwal, P. McCluskey, T. Dishongh, S. Javadpour and R. Mahajan, "Electronic Packaging: Materials and Their Properties", CRC Press, Boca Raton, FL, 1998.

Querk, Michael, and J. Serda, "Semiconductor Manufacturing Technology", Prentice-Hall, Upper Saddle River, NJ, 2001.

Ruska, W. Scot, "Microelectronic Processing: An Introduction to the Manufacture of Integrated Circuits", McGraw-Hill, New York, NY 1987.

CHAPTER 8: REMOVING MATERIALS; FUTURE DEVICES

CHAPTER OVERVIEW

This chapter covers the following methods used to removed materials from IC wafers:

- Developing
- Stripping
- Wet (chemical) etching
- Dry (plasma) etching

This chapter also addresses the topic of what future electronic devices may be like.

8.1: REMOVING MATERIALS

There are not as many ways to remove materials as there to add them, primarily because the existing methods have been quite adequate. As with the section on adding materials, these will be covered starting with the low-tech methods first.

The materials that we need to remove include mainly unhardened photoresist, hardened photoresist, metallization, SiO_2 (doped and undoped), polysilicon, silicon nitride, and polyimide.

8.1.1: Developing

Developing is used to remove the unhardened or softened photoresist from the wafer after exposure in a stepper. Developing is basically a solvent process, in which the developing solution is used to dissolve the soft photoresist and leave the hardened photoresist, creating the desired pattern. The solvent is usually sprayed onto the vertical the wafer, allowing the dissolved photoresist to be carried away as the solvent runs off.

8.1.2: Stripping

Stripping is used to remove the hardened layer of photoresist after it has served its masking function. The vertical wafer is sprayed or immersed in a flow of a chemical which softens the interface between the hardened photoresist and the substrate; it removes the hardened photoresist, and the chemical is filtered to remove the stripped photoresist.

8.1.3: Wet (Chemical) Etching

Wet etching is used to remove the undesired material from those areas previously defined in the photolithography process. The most common material to be etched is oxide, since it is used as the masking material for most selective adding or removing operations. Other chemically etched materials include doped oxide, polysilicon, silicon nitride, and metal conductors. Wet etching is accomplished by immersing the wafer in a flow of a chemical which attacks and dissolves the unwanted material which is not covered with a layer of photoresist or SiO_2.

Wet etching suffers from the serious disadvantage of being an isotropic process. This is depicted in Figure 8.1, which also shows some of the challenges in getting the process just right. As in the case of diffusion (chapter 7), it has problems achieving the high resolution required in most modern ICs, and is falling out of favor for new chip designs, being replaced by dry (plasma) etching.

8.1.4: Dry (Plasma) Etching

Dry etching is used basically as a replacement for wet etching, because it does not suffer from the disadvantage of being isotropic. Most plasma etchants contain either chlorine or fluorine, because they are highly reactive. Figure 8.2 gives a schematic of a dry etch system, commonly known as RIE, for reactive ion etch.

The chlorine or fluorine-containing gas is excited by way of radio-frequency energy to a plasma state, upon which it begins to react with everything in the plasma etching chamber, including the wafer. The plasma is directed preferentially toward the wafer, so most of the reactions take place there. This direction of the plasma causes a highly preferential reaction direction, which is in
the downward direction; very little horizontal reaction (etching) occurs. Figures 8.3 and 8.4 show some features produced by plasma etching; note the vertical sidewalls characteristic of this process. Although plasma etching is more expensive than wet etching, it dominates primarily because of this feature of anisotropy, which permits better process resolution.

Using the right chemical, plasma etching can be done on any of the materials used in IC manufacturing. Indeed, there are some materials for which there is no adequate wet chemical etchant, and only plasma etching will suffice.

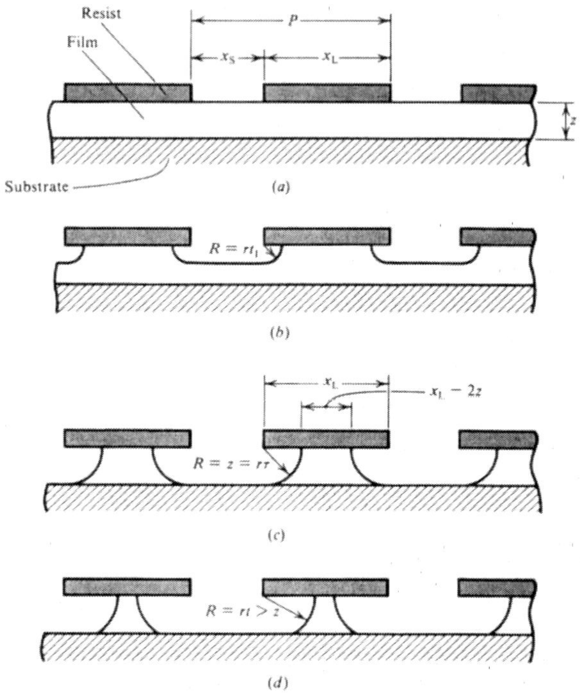

Figure 8.1 Isotropic etching characteristic of wet chemical etching. (a) Unetched, masked film showing parameters to be used, (b) Under-etched (etch time less than ideal), (c) Ideally etched (etch time equal to ideal time), (d) Over etched (etch time greater than ideal). *From Ruska, W. Scot. 1987. Microelectronic Processing: An Introduction to the Manufacture of Integrated Circuits. New York: McGraw-Hill. Courtesy of The McGraw-Hill Companies.*

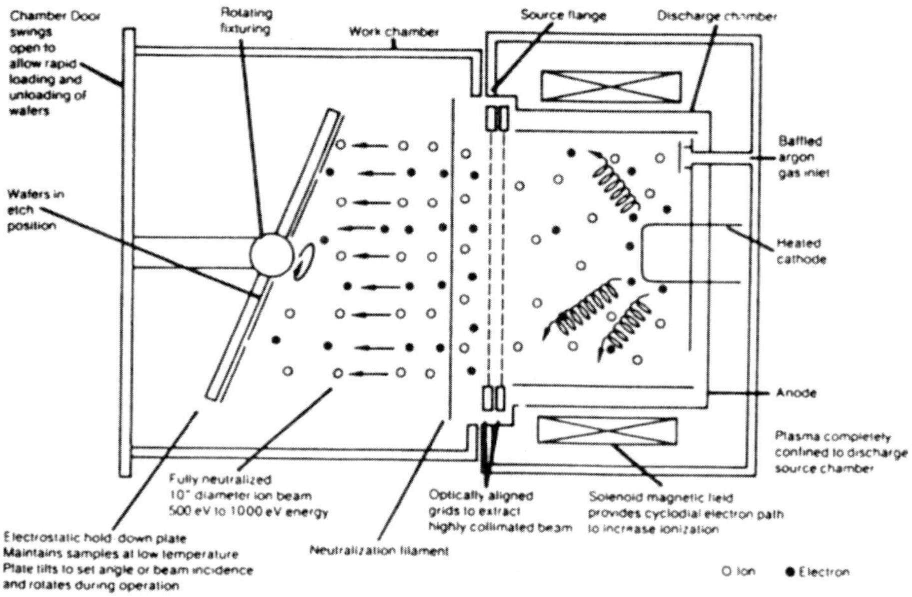

Figure 8.2 A production ion etching (milling) system. *From Ruska, W. Scot. 1987. Microelectronic Processing: An Introduction to the Manufacture of Integrated Circuits. New York: McGraw-Hill. Courtesy of The McGraw-Hill Companies.*

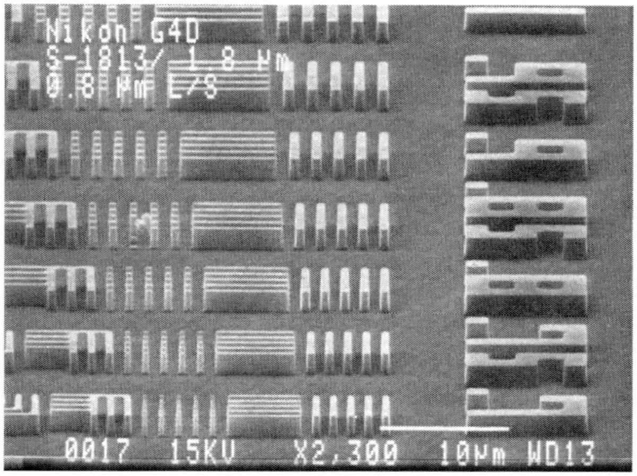

Figure 8.3 Anisotropic results of the ion (plasma) etching process, 0.8 micron. *From Elliott, David. 1989. Integrated Circuit Fabrication Technology. New York: McGraw-Hill. Courtesy of The McGraw-Hill Companies.*

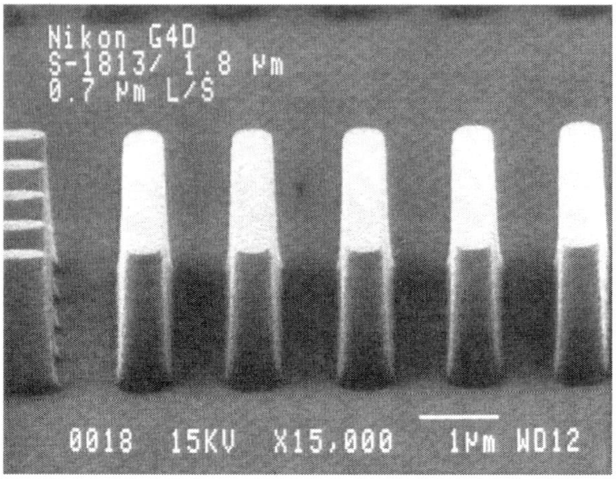

Figure 8.4 Anisotropic results of the ion (plasma) etching process, 0.7 micron. *From Elliott, David. 1989. Integrated Circuit Fabrication Technology. New York: McGraw-Hill. Courtesy of The McGraw-Hill Companies.*

8.2: THE FUTURE OF ELECTRONIC DEVICES

The continual increase in the number of devices per IC has been accomplished primarily by reducing the size of the transistor (and, of course, the other devices on the IC). Obviously, there must eventually come a point where further reduction is not possible. This limitation, often referred to as the fundamental physical limit, has been defined a number of times. Each time these "fundamental limits" have only fallen prey to new methods of getting around the previous "limit." These previous "limits" have therefore not been fundamental limits, but perceived limits. Are we now in a position to state where the true fundamental limit exists? Current literature suggests we are approaching that time, and some authors have tried to identify that limit. It is this author's opinion that the current limitations are twofold: first, the size of the transistor, and second, the practical limitations on continually shrinking geometries. Let's discuss these separately.

8.2.1: Transistor Size

MOSFETs (metal-oxide-semiconductor field-effect transistors), the workhorse transistors of the VLSI era, are not quantum devices; that is, they require large numbers of atoms and electrons to operate. Eventually the transistor could become so small that there would not be enough atoms or electrons to allow the MOSFET to operate as present MOSFETs do. How close are we to that limit? Some ICs in research are being produced using line widths of as little as 10 nm; this means the transistor itself can be as small as 100 nm by 100 nm. Crystalline silicon atoms have a spacing of about 4 Å, which means that transistors made using 100 nm line widths can have approximately 250 by 250 atoms, or 62,500 atoms in the top layer of atoms alone! Clearly, we are not nearing the fundamental quantum limit in terms of transistor area.

However, we are pushing that limit quite severely when it comes to thicknesses of oxides for MOSFETs. Gate oxide thicknesses are presently in the range of 4 nm, which is only about 10 silicon atoms, and this thickness is headed still further down. In this area, we are nearing a very fundamental limit, and quantum effects such as "hot electrons" are a serious problem.

Since 2011, two new variations on the MOSFET, the FinFET and the UTB SOI (ultra-thin body silicon-on-insulator) transistor have begun to dominate the field for very high-resolution processes. More about these can be found in the Ahmed reference listed at the end of this chapter. These new transistor structures have greatly reduced the leakage current in very small transistors, and appear to have solved the problem for another few generations. But a replacement is needed.

Fortunately, the transistor is not our only hope for an electronic workhorse for the future; it should be mentioned here that some companies, working in research for the far future, have demonstrated devices which use quantum effects to switch, or to move or detect single electrons. It is hoped these quantum-effect devices will eventually prove practical. Carbon nanotubes have also shown great potential (see Wikipedia reference).

8.2.2: Practical Limitations

On the practical side, it is readily apparent from the literature that there are several challenges which are becoming increasingly obdurate. The good news is that we seem to continually be able to find adequate short-term solutions, allowing another four or five years, after which some other way must be found. One of these most challenging areas is the wavelength of light used for photolithography. As was discussed in chapter 6, the wavelength of light used for photolithography is directly related to the resolution of the process. Visible light is only a small part of the EMR (electromagnetic radiation) spectrum, and clearly there are frequencies of EMR with much shorter wavelengths than even the EUV presently in use. However, the challenge here is finding practical sources of short-wavelength EMR, and operating on the "light" from these sources such that the resulting EMR can be used in IC photolithography.

Another area where practical limitations are increasingly more challenging is cleanliness. IC manufacturing cleanrooms are already the cleanest places on earth, with air approximately ten thousand times cleaner than on the most pristine mountaintop, and approximately one thousand times cleaner than in the best hospital operating room. Most IC manufacturing facilities have gone to using equipment with mini-cleanroom environments, creating their own super-filtered air within the small area where the wafers are actually processed. Current IC manufacturing cleanrooms are typically Class 10, which for a 30 nm process roughly means that there are fewer than ten particles larger than 30 nm in each cubic foot of air. The mini-environments being installed today qualify as Class 0.01, meaning roughly that there are fewer than ten particles in each 1000 cubic feet of air. But even with all this cleanliness, one of the major limitations to higher yield continues to be foreign material, or contamination, by unwanted molecules in the process environment. IC companies continue to work on this year-in and year-out, with only small advances being made.

One of the major sources of foreign material is the processing chemicals (no chemical is absolutely pure), and the processing equipment itself. Chemicals used to remove materials such as SiO_2, polyimide, metallization, photoresist, etc., are particularly challenging because it is so difficult to filter or otherwise separate from them the materials they have removed. If the removed materials remain in the chemicals, they are only waiting for a place to deposit themselves.

The other major source of foreign material is the processing equipment itself. Although most equipment is lined with stainless steel, even the highest-quality stainless steel is not impervious to the attack of the chemicals and plasmas used in IC processing. Also, the vacuum chambers used for several IC processes are not perfect vacuums, nor can they be perfectly maintained absolutely clean as material is sputtered, evaporated, or plasma etched inside the chamber.

There are other areas in IC manufacturing which are very challenging. The two areas just described represent some of the biggest challenges, but are by no means the only ones.

8.2.3: Future Devices

What will the basic switching and amplifying devices of the year 2035 look like? Many scenarios are presently envisioned, including quantum devices, devices built one atomic layer at a time, carbon nanotubes, biological living devices based on cells, and other more exotic embodiments. Several of these just listed have already been built in the laboratories of research facilities, and research continues. Will the field of silicon FETs eventually run out of steam? Yes, inevitably. But it is this author's opinion that before that grim day comes, there will be other alternatives at the ready, and the current headlong rush into higher and higher levels of technology will not, in the end, slow down, but will continue apace, and perhaps even increase.

CHAPTER SUMMARY

Materials are removed from IC wafers by the following processes:

1. Developing, which is used to dissolve unhardened photoresist. This is basically

a chemical process, and is quite robust and low-cost.

2. Stripping, which is used to remove hardened photoresist. Hardened photoresist is not easy to dissolve, but can be removed from the wafer with application first of a chemical which attacks its bonds to the wafer, then with a spray or scrubbing action.
3. Wet (chemical) etching, which is used to remove a material which is not protected by hardened photoresist or SiO_2. The wet chemical used must be one which will dissolve the material which must be removed, so the process is very selective, but is isotropic.
4. Dry (plasma) etching, which is used to remove almost any material. Using the right chemical in the chamber, the plasma excites it to the reactive state, and it reacts with and removes most materials except for photoresists. It is highly versatile and, but requires a high vacuum.

FOCUS QUESTIONS

1. Two of the processes discussed in this chapter are applicable almost exclusively to photoresist. Which are they?
2. In wet etching, what problem(s) might result if the chemical being used is not entirely pure?
3. In wet etching, what problem(s) might result if the process is not performed long enough?
4. Why is dry etching more versatile than wet etching?
5. Why has dry etching virtually replaced wet etching?
6. In what dimension of the transistor are we nearing some fundamental quantum limits?
7. What are some of the practical limits presently being encountered in making ICs?

FOR ADDITIONAL READING

Ahmed, Khaled & Klaus Schuegraf, "Transistor Wars: Rival Architectures Face Off in a Bid to Keep Moore's Law Alive", *IEEE Spectrum*, Nov 2011, 50-55.

Elliott, David J., "Integrated Circuit Fabrication Technology", Chapter 1, McGraw-Hill, New York, NY 1989.

Ghandi, Sorab K., "VLSI Fabrication Principles: Silicon and Gallium Arsenide", John Wiley & Sons, New York, NY, 1994.

Harper, Charles A., and R.M. Sampson, "Electronic Materials & Processes Handbook", McGraw-Hill, New York, NY, 1994.

Pecht, Michael, R. Agarwal, P. McCluskey, T. Dishongh, S. Javadpour and R.

Mahajan, "Electronic Packaging: Materials and Their Properties", CRC Press, Boca Raton, FL, 1998.
Quirk, Michael, and J. Serda, "Semiconductor Manufacturing Technology", Prentice-Hall, Upper Saddle River, NJ, 2001.
Ruska, W. Scot, "Microelectronic Processing: An Introduction to the Manufacture of Integrated Circuits", McGraw-Hill, New York, NY 1987.
Wikipedia, "Carbon nanotube field-effect transistor", April 2011.

CHAPTER 9: INTEGRATED CIRCUIT PACKAGES: PROCESSES AND MATERIALS (LEVEL 1)

CHAPTER OVERVIEW

This chapter covers the following concepts:

- Some chips are used in bare-die form (more in next chapter)
- The parts of the basic IC package
- How chips are electrically and mechanically attached to their packages
- IC packaging materials and some of the properties

9.1: INTRODUCTION

Level 0 of electronic manufacturing (IC chips), covered in the preceding three chapters, is easily the most complicated and sophisticated field of electronic manufacturing. This chapter will cover level one, which also has its share of challenges and continual advances. Much of the miniaturization that has happened over the past few decades has been due to advances at this level, and there is much improvement yet to come.

9.2: IC PACKAGING (LEVEL 1)

Before we move on to level 1, we should acknowledge that in some cases, the bare chips (level 0) are the final product. Companies which manufacture hybrid ICs or multi-chip modules (MCMs) buy the bare chips after they have been tested at the wafer level, and incorporate them in their modules. This will be covered in more detail in the next chapter. Also, manufacturers of "smart" cards (credit-card sized with on-board intelligence and storage) use bare die in this application. And finally, there are manufacturers that put bare die on the PWB, in processes known as chip-on-board (COB) and direct-chip attach (DCA).

However, despite these and other emerging applications for bare die, most ICs are put into packages before they go to the end user. A detailed discussion of the function of these packages is found in chapter 2. In this chapter we will discuss the process by which this packaging is done and the materials commonly used for IC packages.

9.3: THE IC PACKAGING PROCESS

IC packages consist of four basic parts (see Figure 9.1).

1. the silicon IC chip
2. the lead frame or wiring substrate
3. the connections from the IC chip to the lead frame or wiring substrate
4. the mold compound or encapsulation material

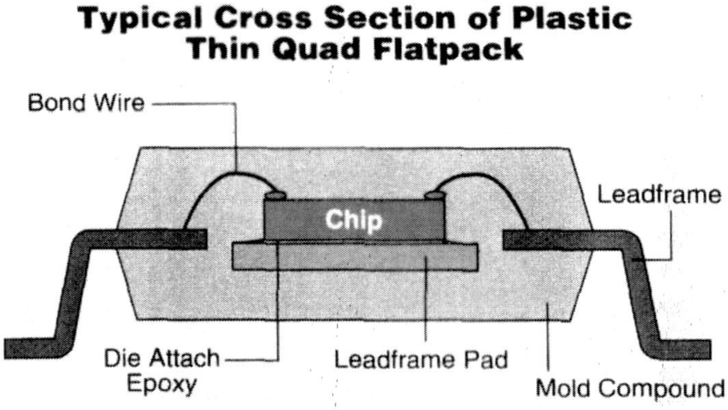

Figure 9.1 A cross section of a typical IC package. From Howard Markstein. October, 1995. Electronics Production & Packaging. Used by permission.

Most IC packages can be categorized in the following ways:

1. whether the leads are on the periphery of the package (lead frame or TAB – tape automated bonding) or in an array beneath the package (grid-type packages). In Figure 9.2, only the PGA (pin-grid array) is an example of an array package; the other five shown have the leads along the periphery.
2. whether the chip is connected to the leads via the C4 (controlled collapse chip connect) process, via the TAB process, or via wire bonds. In Figure 9.2, the PGA uses the C4 method, the Flatpack often uses the TAB process, and the others use wire bonds.
3. whether the package enclosure material is either metal, ceramic, or plastic. In Figure 9.2, none are metal, the LCC (leadless chip carrier), PGA and DIP (dual in-line package) are ceramic, and the SOIC (small-outline integrated circuit), PLCC (plastic leaded chip carrier), and Flatpack are plastic.
4. by the type of material the connection substrate is made from, whether they are TAB lead frames, A42 (Alloy 42) lead frames, organic wiring substrates (such as epoxy/fiberglass), or ceramic wiring substrates (usually Al_2O_3). In Figure

9.2, the Flatpack is typically a TAB frame, the PGA is ceramic, and the other four use an A42 lead frame.

The silicon chip has been discussed in detail in the preceding chapters, so let's begin this section by discussing the lead frame or wiring substrate.

9.3.1: A42 Lead Frames and Wirebonding

A42 lead frames are typically made from alloy 42, commonly referred to simply as A42. This is a mixture of nickel and iron, designed to have essentially the same coefficient of thermal expansion (CTE) as the silicon itself.

Most peripheral-lead IC packages, such as the DIP, SOIC, PLCC, and Flatpack (see Figure 9.2) use a metal lead frame onto which the chip is bonded

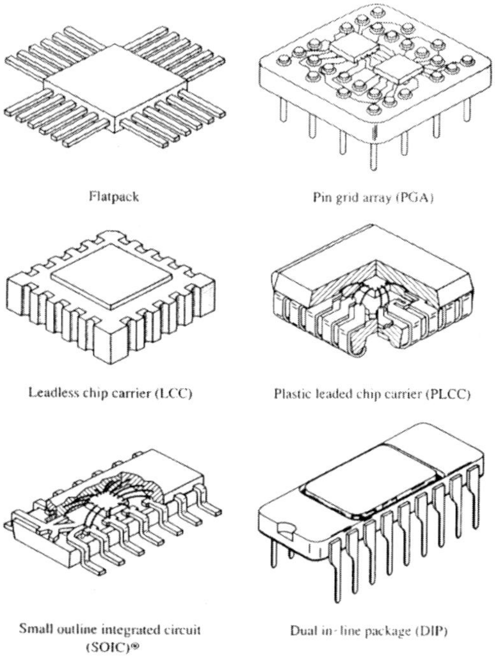

Figure 9.2 Typical IC packages. *From Seraphim, Donald P., Ronald Lasky, and Che-Yu Li. 1989. Principles of Electronic Packaging. New York: McGraw-Hill. Courtesy of The McGraw-Hill Companies.*

with an adhesive. The adhesive is dispensed onto the lead frame or onto the back side of the chip, then the chip is placed face up on the lead frame and the adhesive is cured by either elevated temperature or exposure to UV light.

The next step in packaging a chip is to make the connections from the IC chip to the lead frame, which in this package type is done by wirebonding. The wire used is either gold or aluminum, depending on the requirements of the package, and is only about 1 mil (25 μm) in diameter. Gold wire is usually bonded by thermocompression or thermosonic bonding; aluminum wire is usually bonded by ultrasonic or thermosonic bonding. The process by which these bonds are made is shown in Figure 9.3.

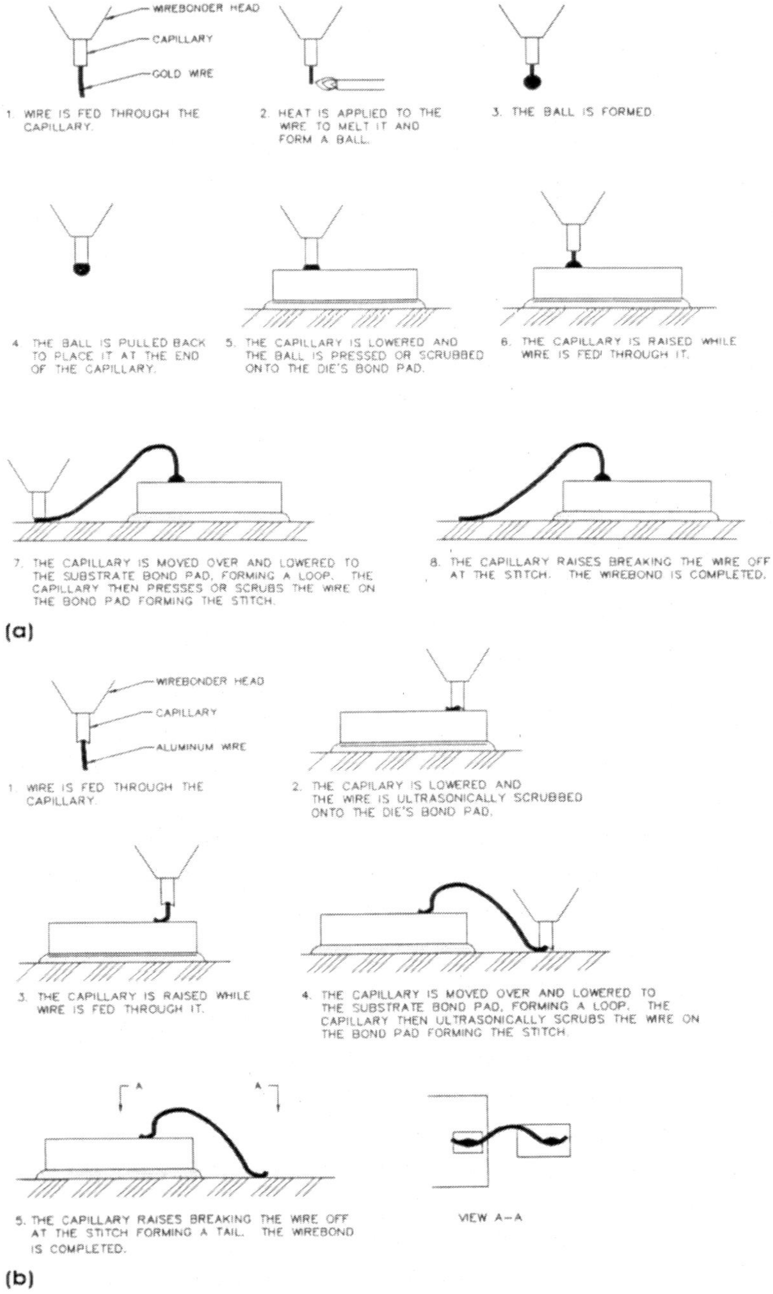

Figure 9.3
Formation of (a) a ball wire bond and (b) a wedge bond. *Reprinted from Handbook of electronic Packaging Design, p. 164 by courtesy of Marcel Dekker, Inc.*

An example of these bonds is shown in Figure 9.4.

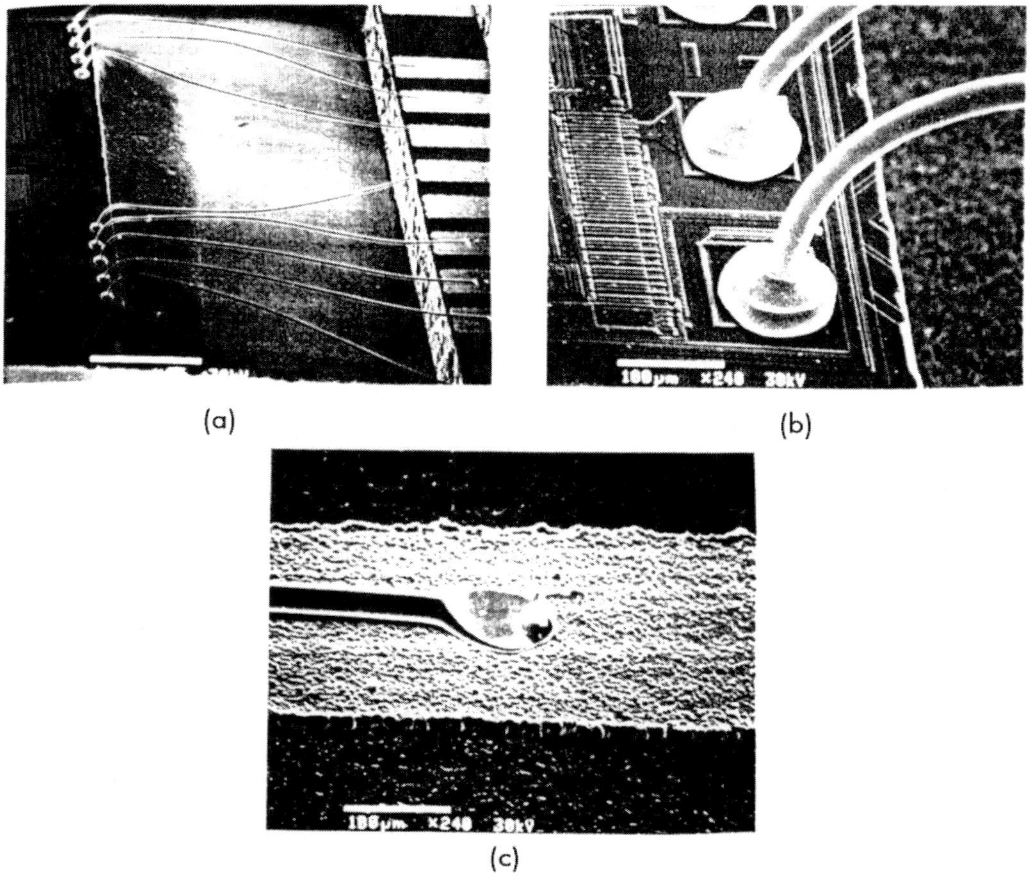

Figure 9.4 SEM images of wire bonds. (a) An SEM image of a ball-wedge type of wire bond, (b) A higher magnification of the ball bond at the chip pad, (c) A higher magnification of the wedge bond at the chip carrier pad. *From Seraphim, Donald P., Ronald Lasky, and Che0Yu Li. 1989. Principles of Electronic Packaging. New York: McGraw-Hill. Courtesy of The McGraw-Hill Companies.*

Thermocompression bonding uses both an elevated temperature and the energy of compression to form a permanent metallurgical bond between the wire and the metallized pads of the IC chip. The wire is dispensed from a heated tip, which is then pressed onto the IC chip pad. The compression also provides a slight wiping action between the wire and the pad, which removes any thin oxide layer which may have formed on either.

Ultrasonic bonding uses a dispensing tip that is vibrated at an ultrasonic frequency (usually around 80 kHz). When the wire from this vibrating tip is brought into contact with the IC chip pad, it provides a scrubbing action, which both removes any oxides which may have formed, and imparts heat to the wire and pad by the friction of the scrubbing action. Again, a permanent metallurgical bond is formed.

Thermosonic bonding uses a combination of the two; the dispensing tip is both heated and vibrated at an ultrasonic frequency. As with the other methods, a permanent metallurgical bond is formed.

Wire bonds can be made at very high speed using modern automated equipment; anywhere from five to twenty bonds per second can be formed, depending on the distance between the chip pad and the lead frame contact point. However, even at these speeds, one of the biggest disadvantages of wirebond connections is that they must be made one at a time. This means that it takes longer than a method which makes all the connections at once; and it means that the reliability of the connections as a whole is better. Two methods which make all the connections at once are TAB and C4; these are covered in the next two sections.

9.3.2: Tape Automated Bonding (TAB)

An example of a TAB package is shown in Figure 9.5; a TAB lead bump is depicted in Figure 9.6. The wiring connections for this package method are made on polyimide substrates by the same type of etching and photolithography processes by which printed wiring boards are made. Basically, a thin sheet of copper is laminated to a thin sheet of polyimide. The copper is then covered with a photoresist, and the photoresist is exposed and developed. Then the copper is etched where the photoresist has been removed by the developing process. These polyimide/copper sandwiches are made in long tapes, from which the name is derived, and can be mass produced at low cost. After the copper is etched, the leads are "bumped," or else the pads on the chip are "bumped." The chip is then bonded to the TAB by heating all of the lead ends of the TAB as it lies on top of the chip.

Figure 9.5 The inner leads of a TAB package, where the chip is attached. *From Harper, Charles. 1997. Electronic Packaging and interconnection Handbook. New York: McGraw-Hill. Courtesy of The McGraw-Hill Companies.*

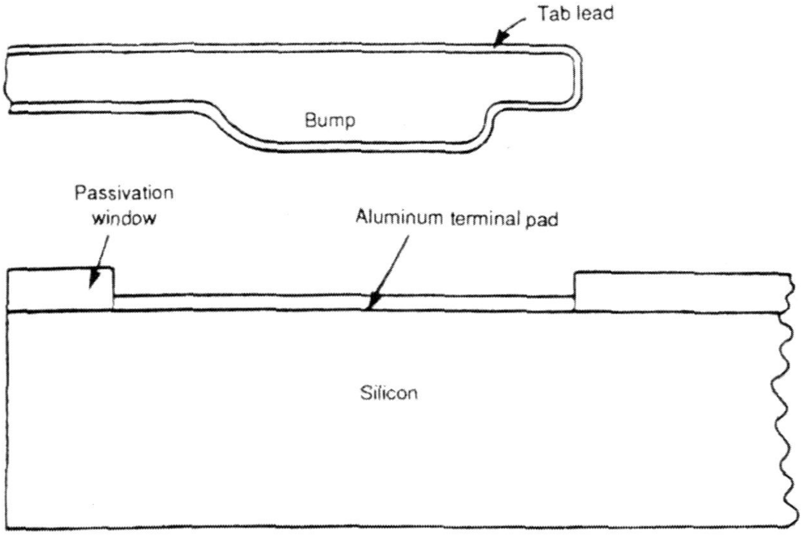

Figure 9.6 Geometry of a bumped lead for a TAB package. *From Harper, Charles. 1997. Electronic Packaging and interconnection Handbook. New York: McGraw-Hill. Courtesy of The McGraw-Hill Companies.*

The advantages of the TAB method are that all the connections are made at once, making the process reliable and fast. Also, it is inexpensive for large volumes, since the TAB wire frames are not expensive in of themselves. The disadvantage is primarily that each different chip must have its own TAB wire frame custom-designed for it, making it impractical for low volumes.

9.3.3: Wiring Substrates and C4

Packages that have all the leads underneath the package, such as PGA (pin grid array) and BGA (ball grid array), do so using the C4 connection method and a wiring substrate. The wiring substrate, usually using several layers of connections, provides the wiring from the pad sites of the IC chip to the pins or balls of the array package. The wiring substrate is usually made either from ceramic material (covered in more detail in section 9.4) or from plastic (organic) material (covered in more detail in section 9.5). Before the chip can be connected to the wiring substrate, it must first be "bumped," a process by which small solder balls are formed on the pads of the IC chip (see Figure 9.7). To connect and bond the chip to the wiring substrate, the chip is flipped face-down (this process is also commonly known as the "flip-chip" method), the small solder balls are aligned with the wiring substrate, the chip is lowered onto the wiring substrate, and the chip is heated under a very carefully controlled thermal process. This heating leads to a controlled collapse, and metallurgical bonding occurs between the IC chip pad bumps and the wiring substrate, which give the process its name, controlled-collapse chip connect (C4).

Like the TAB process, one advantage of the C4 process is that all the connections are made at once. Also, the alignment between the chip and the wiring substrate does not have to be highly accurate, because the heating and collapsing process of the solder balls causes a liquid tension force which tends to automatically center the chip pads over the wiring substrate. This makes for a very high-reliability chip connection method. The disadvantage of this process, like with the TAB process, is that each chip must have its own, custom wire substrate formed, which is not the case for the wirebonding method.

9.3.4: IC Package Lead Pitch

One of the most significant challenges in IC packaging is that of I/Os. Most high-performance (meaning high-frequency or high-density) ICs need lots of I/Os.

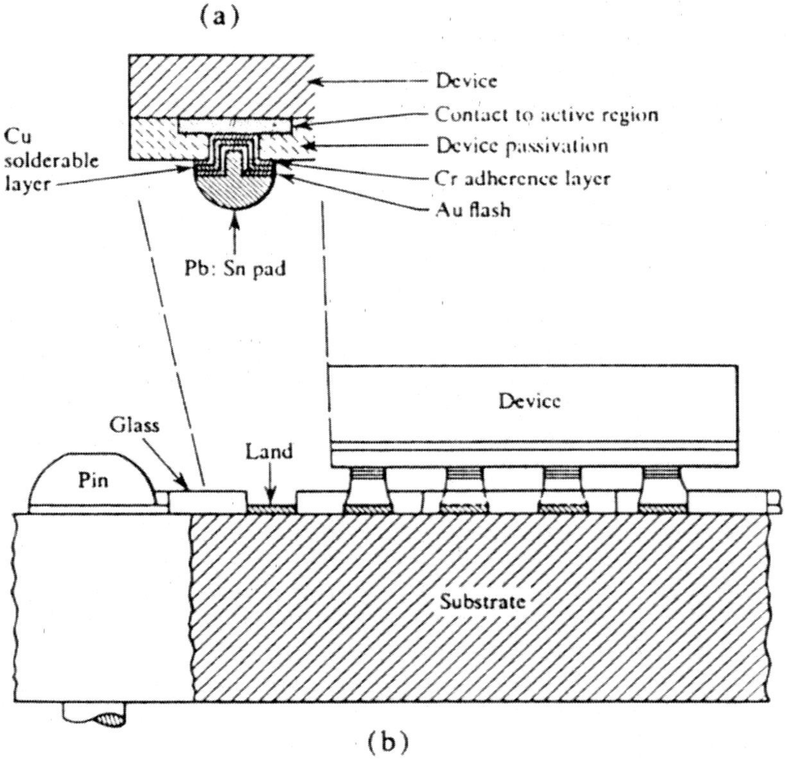

Figure 9.7 The C4 process. (a) Side view of a device (b) Side view of a device on a substrate (dam method). *From Tummala, Rao R. and Eugene J. Rymaszewski. 1989. Microelectronics Packaging Handbook. New York: Kluner Academic Publishers.*

Today's IC packages have as many as 3000 I/Os, and plans are that this number needs to grow to over 8,000 by the year 2020. With the need for so many I/Os, the responsibility for meeting this need falls to the IC package. There are two basic ways to arrange the I/Os on IC packages: leads on the periphery of the package, or in an array of pins or balls beneath the package. The advantage of the array is significant; with so much more area over which to distribute the pins or balls, they can be much further apart, which makes them mechanically much easier to make and handle. However, this is not the case with packages with peripheral leads. In order to pack a large number of leads on the periphery, the leads must be extremely close together. The distance between the centers of adjacent leads, as shown in Figure 2.3, is termed the lead pitch. As the pitch grows finer, the leads become thinner, making the leads very fragile mechanically; they are easily bent. The pads on the PWB must also be closer, which makes the PWB more expensive, and the process of attaching this type of IC package to the PWB by reflow soldering (see chapter 12) becomes a much bigger challenge.

Today's array-lead packages have pitches as fine as 25 mils, which is fine; but periphery-lead packages now go as fine as 12 mils, and this is becoming a major manufacturing challenge because the leads are so fine and easily bent.

9.4: IC PACKAGING MATERIALS

IC packages come in two basic kinds: hermetic and non-hermetic. Hermeticity was described previously in chapter 2. Hermetic packages are generally made from either metals or ceramics. Non-hermetic packages are generally made from an epoxy-phenolic plastic known as novolac. Figure 9.8 shows the rates of moisture permeation of several packaging materials. From this figure, it can be seen that plastic packages (novolac) are basically ineffective at preventing moisture permeation. Ceramic packages (in the same range as glasses) are effective for many years; metal packages are effective for many decades. As was discussed in chapter 2, the disadvantage of hermetic packages is that they are more expensive. However, hermetic packages do protect very effectively against two of the major causes of stress to electronic circuits: moisture and dust.

Metal packages, although more expensive than plastic packages, have another advantage: they generally can handle higher wattages than plastic or ceramic packages, because the material of the package has a higher thermal conductivity. For this reason, metal packages are generally reserved for devices which dissipate relatively high amounts of power (>5W). Exceptions exist, such as the ubiquitous TO-5 package, which is metal yet small, and so can only dissipate small amounts of heat.

However, because it is metal, it can readily accept heat sinks which allow it to dissipate much more than it otherwise could.

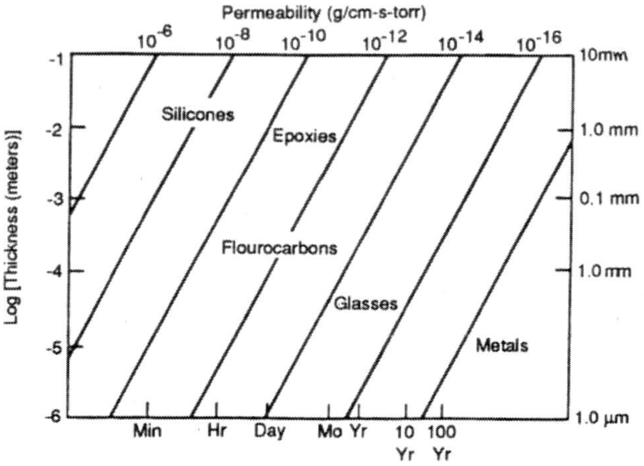

Figure 9.8 The time required to permeate various packaging and sealing materials. *Reprinted from Handbook of Electronic Packaging Design, p. 164, by courtesy of Marcel Dekker.*

Tables 9.1 and 9.2 show the thermal conductivity for several IC packaging materials.

Material	Dielectric Constant at 1 MHz	Thermal Conductivity (W/m·K)	Coefficient of Thermal Expansion (ppm/°C)
Alpha quartz (SiO_2)	4.8	2	6.0
Alumina, low purity (<99% pure)	9.0	20	7.0
Alumina, high purity (≥99% pure)	10.0	30	6.7
Aluminum nitride (AlN)	9.2	200	4.3
Beryllium oxide (BeO)	7.0	250	6.8
Boron nitride (cubic)	4.1	1300	3.8
Boron nitride (hexagonal)	4.1	60	3.8
Diamond	5.6	2000	1.2
Forsterite ($2MgO{\cdot}SiO_2$)	6.2	3.3	9.8
Fused quartz (SiO_2)	3.8	1.6	0.5
Silicon nitride (Si_3N_4)	7.0	30	2.8
Spinel ($MgO{\cdot}Al_2O_3$)	8.3	15	7.0
Steatite ($MgO{\cdot}SiO_2$)	5.7	2.5	7.2

Table 9.1 Properties of several ceramic IC packaging materials.

Ceramic packages are also hermetic, and have better thermal conductivity than plastic packages. Ceramic also has a lower CTE than plastic, and its CTE is more closely matched with that of silicon (2.3 ppm/°C). Table 9.1 lists several properties of common ceramic materials used for IC packaging.

Plastic is the most common material used for making IC packages, which is somewhat surprising when you consider that it has the poorest engineering characteristics. It is quite permeable, is has a relatively high CTE (≈40 ppm/°C), its CTE is much higher than that of silicon (2.3 ppm/°C), and its thermal conductivity is much poorer (≈3.4 W/m·K) than that of metals or ceramics. However, it is quite adequate in toughness, and is heavily used, primarily because of its low cost. As mentioned previously, the most common plastic used in IC packages is often called novolac, and is a blend of epoxy and phenolic resins; many formulations exist. Its dielectric constant is about 6.0.

Material	Dielectric Constant at 1 MHz	Thermal Conductivity (W/m·K)	Coefficient of Thermal Expansion (ppm/°C)
FR4	4.5	1.0	16
Novolac	6.0	1.5	40
Polyimide	3.5	.15	50

Table 9.2 Properties of several plastic IC packaging materials.

Some of the most recent IC packages use printed wiring board material, commonly known as FR4 (for fire-retardant type 4), for the wiring substrate, as shown in Table 9.2. FR4 is made from a fiberglass cloth which is impregnated with epoxy resin, and so has properties which are a combination of both the fiberglass and the epoxy. Also sometimes used is polyimide, as mentioned in section 9.3.2, primarily in TAB packages. Some properties of these materials are summarized in Table 9.2.

CHAPTER SUMMARY

Although bare die are being used in an increasing number of applications, the majority of today's chips are packaged before being shipped from the IC manufacturing company. These packages consist of the chip, the leadframe, the connections from the chip to the leadframe, and the packaging material. Most leadframes are made from A42, an alloy of nickel and iron.

Most chips are electrically connected to the leadframe using aluminum or gold wires, and bonded to the die and the leadframe via one of three wirebonding methods: thermocompression, ultrasonic, or thermosonic. Other methods used include tape-automated bonding (TAB) and the C4 (controlled-collapse chip connect) process.

Lead pitches have come down significantly over the years, greatly adding to the difficulty of making PWB assemblies. PGA and BGA packages are a great relief to this trend, although not without problems of their own.

There are several desirable characteristics for IC packaging materials, and no single material is ideal for all applications. Plastic packages are, by far, the most common, although they are not hermetic. Hermetic packages include ceramic and metal packages, but they are much more expensive.

FOCUS QUESTIONS

1. What are some of the applications for which bare-die are presently being used?
2. Which IC package category (peripheral-leads or grid-array leads) is inherently capable of more pins per unit area? Why?
3. Which electrical connection method (C4, TAB, or wirebonding) must be done one lead at a time?
4. Which electrical connection method (C4, TAB, or wirebonding) features a property sometimes called self-centering, thus making the alignment a simple matter?
5. Which electrical connection method (C4, TAB, or wirebonding) is capable of connecting to pads anywhere on the face if the die, instead of only on the edges?
6. Which two of the electrical connection methods (C4, TAB, or wirebonding) require "bumping" of the chip (or package) first?
7. What is one of the significant challenges of fine-pitch (≤25 mils) peripheral-lead packages today?
8. Which type of packaging material(s) would you expect to find in life-critical and high-reliability applications?
9. The plastic used for most IC packages is quite poor in its engineering performance characteristics. Why is it still used so much?

FOR ADDITIONAL READING

Blackwell, Glenn R., "The Electronic Packaging Handbook", Chapter 3, CRC Press, Boca Raton, FL, 1999.

Harper, Charles A., "Electronic Packaging and Interconnection Handbook", Chapters 1, 10, McGraw-Hill, New York, NY, 1997.

Harper, Charles A., and R.M. Sampson, "Electronic Materials & Processes

Handbook", McGraw-Hill, New York, NY, 1994.
Lau, John H., C.P. Wong, J.L. Prince, and W. Nakayama, "Electronic Packaging: Design, Materials, Process, and Reliability", Chapters 1, 2, McGray-Hill, New York, NY 1998.
Matisoff, Bernard S., "Handbook of Electronics Packaging Design and Engineering", Chapter 1, Van Nostrand Reinhold, New York, NY, 1990.
Pecht, Michael, "Handbook of Electronic Package Design", Chapter 12, Marcel Dekker, New York, NY, 1991.
Pecht, Michael, R. Agarwal, P. McCluskey, T. Dishongh, S. Javadpour and R. Mahajan, "Electronic Packaging: Materials and Their Properties", CRC Press, Boca Raton, FL, 1998.
Quirk, Michael, and J. Serda, "Semiconductor Manufacturing Technology", Chapter 20, Prentice-Hall, Upper Saddle River, NJ, 2001.
Seraphim, Donald P., R.C. Lasky, and C. Li, "Principles of Electronic Packaging", Chapters 2, 10, 11, McGraw-Hill, New York, NY, 1989.
Tummala, Rao R. and Eugene J. Rymaszewski, "Microelectronics Packaging Handbook", Van Nostrand Reinhold, New York, NY, 1989.

CHAPTER 10: HYBRIDS, MCMs, AND PWBs

CHAPTER OVERVIEW

This chapter covers the following concepts:

- Hybrids and MCMs are known as level 1½
- The advantages and disadvantages of hybrids and MCMs
- The substrates and devices used for hybrids and MCMs
- The process of making PWBs
- What PWBs do for us
- What the desirable characteristics are for PWB materials
- The properties of many PWB materials
- Some of the terminology common to PWBs

10.1: HYBRID ICs AND MCMs (LEVEL 1½)

IC chips are level 0; IC packages are level 1. Level 2 is printed wiring boards, which puts hybrid ICs and MCMs right between levels 1 and 2, so they have commonly been referred to as level 1½. MCMs and hybrids are very similar to each other, but there is a difference; an MCM is a multi-chip module, meaning that it has more than one IC chip; a hybrid IC, by definition, has more than one type of component (more than just IC chips). When we put more than one IC chip on a single substrate, then package that substrate in preparation for putting it on a printed wiring board, we have made an MCM. An example of one is seen in Figure 10.1. When we put an IC chip together on a common substrate with a chip resistor or capacitor or a thick film resistor, and then package that substrate in preparation for putting it on a printed wiring board, we have made a hybrid IC. An example of a hybrid is seen in Figure 10.2.

10.1.1: Advantages and Disadvantages

Why would we be interested in making MCMs and hybrid ICs? Well, they have several advantages, including the following:

1. They allow more chips and/or components to be placed together in a smaller area. In a word, they have greater density.
2. They use less packaging material, because the chips are not packaged before placing on the substrate. This means that the equivalent function in an MCM or

hybrid package will weigh less.

Figure 10.1 An MCM with 20 IC chips (die). *From Advanced Packaging, Jan 1998, p. 22. Reprinted by permission.*

Figure 10.2 A hybrid IC package with 3 chips (die), 16 chip resistors and capacitors, and 20 film resistors. *Reprinted from Handbook of Electronic Package Design, p. 196, by Courtesy of Marcel Dekker, Inc.*

3. They allow more chips and/or components to be placed together in a smaller area. In a word, they have greater density.
4. They use less packaging material, because the chips are not packaged before placing on the substrate. This means that the equivalent function in an MCM or hybrid package will weigh less.
5. Advantages one and two lead to another advantage, after we reach high volumes: lower cost. The reason we do not realize this advantage until we reach high volumes is because of the NRE (non-recurring engineering) costs associated with the design, development and test of the MCM or hybrid substrate and package.
6. Advantage #1 leads to another advantage: higher performance. Because the devices are closer together, there is less parasitic capacitance and less distance for the signals to travel, which leads to higher frequency performance.

As we well know by now, there is rarely an engineering alternative that does not have its tradeoffs, or disadvantages. In the case of MCMs and hybrids, these tradeoffs include:

1. Relatively high NRE costs, as mentioned in #3 above. However, at high volumes these costs can be recovered.
2. Lack of familiarity. Most engineers have not designed MCMs or hybrids, and most IC manufacturing companies do not make them. Therefore, when undertaking to make one of these, the path is less familiar and often takes longer than expected.
3. Changes in IC chips and other components. If any of the IC chips or other components used in the MCM or hybrid undergo any significant change, be it electrical or physical, it usually impacts the design of the MCM or hybrid. This can often be problematic.
4. Known good die (KGD). IC chips, after they are packaged in modules, usually undergo much more thorough testing than they do at the chip level. They also often undergo stress testing such as burn-in and mechanical tests. Generally, bare IC chips have not been tested as thoroughly, which means that the IC chips used to make up MCMs and hybrids are not as high quality as the packaged versions. This is especially a challenge when the MCM or hybrid uses many IC chips, because there is a significant probability that one or more of the chips will be bad, and the costly options then are scrapping the entire module, or reworking it, involving removal of the bad chip(s), preparation of the site where the chip was removed, attaching a new chip, and then testing the entire module again. Much effort has been poured into this problem, and great progress has been reported, but the KGD challenge still exists.

10.1.2: Substrates

The wiring substrates used for MCMs and hybrids are generally made from the same materials as those used for IC packages. Ceramic substrates are very common, being on the high-performance end; FR4 and polyimide substrates are also sometimes used. The reader is referred to chapter 9 to review the properties of these materials.

Multiple wiring layers are used, numbering from four to as high as forty, depending on the complexity of the MCM or hybrid. Substrates with more than ten wiring layers are almost always made of ceramics (usually alumina).

10.1.3: Devices for Hybrid ICs

As the name implies, hybrid ICs contain a mixture of devices. They always contain at least one IC, and usually two or more. They also usually have a few chip resistors or capacitors, which were described in chapters 3 and 4, as well as a few thin or thick-film resistors.

Thick-film resistors are made by the same process described for network resistors in chapter 3. The resistive ink is deposited in the desired places on the wiring substrate, the ink is cured, and the resistors are trimmed. Thick-film resistors are cheaper and more reliable than chip resistors, but are generally not available in a wide range of values (greater than 100:1) on the same substrate.

10.2: PRINTED WIRING BOARDS: FABRICATION

Printed wiring boards (PWBs), often also called printed circuit boards (PCBs), have been around for over sixty years, and while they have evolved to meet the increasing demands placed on them by evolving circuit design, they remain very similar to the early ones not so long ago.

The process of making PWBs is one of the processes from which came many ideas for making early ICs. Specifically, the idea of using a photoresist and the process of photolithography for defining areas to be etched away, has been used for many years in the manufacture of PWBs.

Figure 10.3 depicts the structure of a multi-layer PWB. We will go through the procedure of making one of the inner cores, after which we will describe the procedure of making the entire assembly. The process described is that used to make PWBs from FR4, which are by far the most common type of PWBs in the industry.

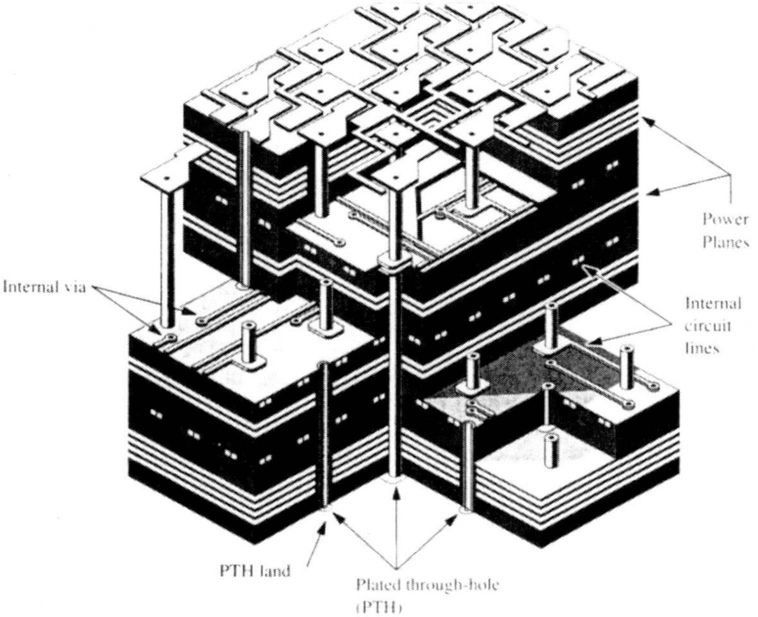

Figure 10.3 The internal structure of a multilayer PWB. *From Seraphim, Donald P., Ronald Lasky, and Che-Yu Li. 1989. Principles of Electronic Packaging. New York: McGraw-Hill. Courtesy of The McGraw-Hill Companies.*

First, a thin sheet of copper foil is laid down on a clean surface. Next, the required number of layers of FR4 prepreg sheets are laid on top of the copper foil, as shown in Figure 10.4. Prepreg will be described later in section 10.4. Finally, on top of this stack of copper foil and prepreg sheets, we lay another thin sheet of copper foil. This sandwich is then placed in a heated press for a specific amount of time, then slowly cooled. The heat and pressure cause the resin in the prepreg to flow and adhere to the sheets of copper foil and to each other; when the sandwich has cooled, the final product is a flat, stiff layer of FR4 with copper foil sheets permanently attached to each side.

Figure 10.4 A diagram showing copper foil sheet and FR-4 sheets used to make a core for PWBs.

Each side of the FR4 sandwich is then coated with photoresist, the photoresist is partially dried or cured (hardened), and the photoresist is imaged through a mask defining the areas of copper to be removed. The photoresist is developed, which removes the unhardened photoresist and exposes the bare copper in those areas defined by the mask. Then the sandwich is etched in a chemical that dissolves copper, such as sodium persulfate, ammonium persulfate, or ferric chloride. After the unwanted copper has been etched, the hardened photoresist is removed in a stripping operation that involves the immersion or spraying of the sandwich with a chemical that softens the hardened photoresist, allowing it to be removed. The final product of this stripping step is a sandwich which has connections in copper on each side; this is called a core. A multi-layer PWB is made up of multiple cores, along with the final outer layers, as shown in Figure 10.5.

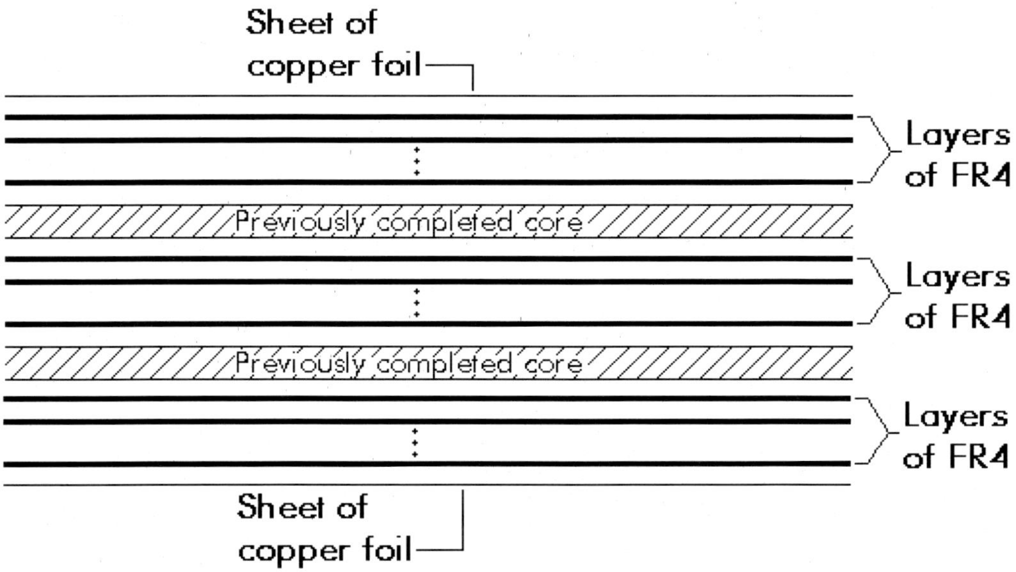

Figure 10.5 A cross section of a six-layer PWB with two two-layer cores, two outer layers, and several layers of FR-4.

After the entire sandwich of cores and outer layers has been processed such that the outer layers have also been patterned, the entire assembly is ready to be drilled. Boards are stacked two to four deep in an NC drilling machine (depending on their thickness), and the holes are drilled using very high-speed (over 80,000 rpm is common), dedicated drilling machines that can drill up to 8 holes/second in the stack. Holes are drilled where there will need to be metal between all the layers to make connections; after the holes are drilled, the barrel of the hole is the substrate material (FR4). In order to make the connections between the layers, the hole needs to be coated with conductive material. This is done by first removing any resin smear that may have covered inner-layer pads that need to connect to the hole. Then the board is immersed in an electroless copper deposition tank, where a copper-laden chemical deposits copper on everything immersed into it. After a thin layer has been deposited on the inside of the hole, the thickness of this layer is increased many times through electro-deposition of copper. Electro-deposition is basically an electroplating process, in which the PWB becomes the anode of the electroplating bath, picking up copper ions as current passes through the electroplating solution.

At the completion of the electroplating process, the PWB is basically complete as far as the wiring patterns are concerned. There are other operations which must follow to make the board complete, which may include the following:

1. A layer of solder-resist is either screen-printed or coated and then photo-imaged on the PWB. This solder-resist or solder mask prevents solder from forming shorts where it is not wanted.
2. The outer-layer pads are covered with a solderability preservative. Between the time the PWB is manufactured and the time it is populated with components and soldered, the copper has a tendency to oxidize, which makes it very difficult to solder. Preservation of solderability is accomplished by either hot-air solder leveling (HASL), or through the application of an organic solderability preservative (OSP). HASL is depicted in Figure 10.6, and involves immersing the PWB in a pot of molten solder, and while removing the PWB, hitting it with blasts of hot air to remove excess solder from the holes and outer surfaces. The application of an OSP is accomplished either by immersion or by spray coating the board.
3. Component identification and location legends are screen-printed onto the board, typically using a white ink. The mask or screen for this process has been photo-defined.
4. Edge connections on the board must be specially treated. First, a layer of nickel is coated on top of the copper of the edge connections through electroless deposition. Nickel forms a hard underlayer for the gold which is to come, and prevents the gold atoms from migrating into the copper. Finally, a thin layer of gold is deposited through electroless deposition. The gold forms the contact for the connectors, since it retains its excellent conductivity and will not oxidize over time.
5. Bevels are cut on the PWB edge connectors to make the PWB easier to insert into the mother board.

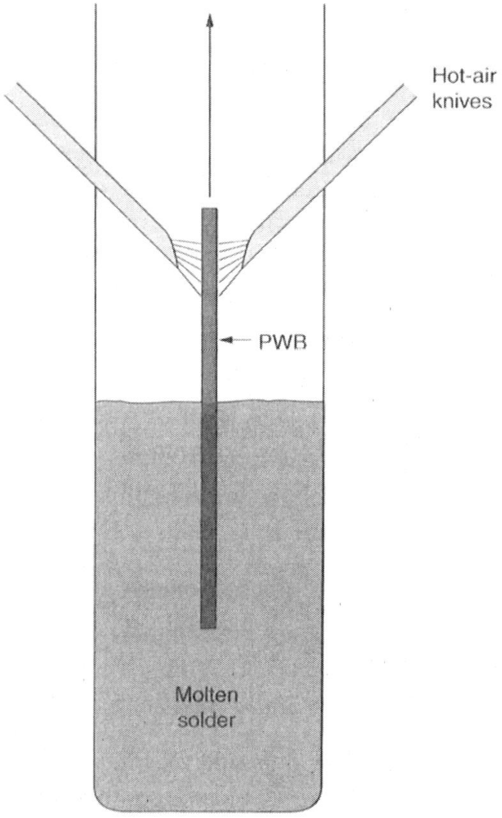

Figure 10.6 The HASL process.

10.3: THE FUNCTIONS OF PWBs

Printed wiring boards accomplish several functions in modern electronic products, including at least the following:

1. They provide a means of mechanically fixing the components in place.
2. They provide the electrical connections for signals, power and ground.
3. They provide electrical separation between pins, components, and connections.
4. They help remove the heat generated by the components they carry.
5. They provide for the testability of the of the assembled PWB.

The degree to which a PWB accomplishes these and other functions determines the performance of that PWB. There have been many types of PWBs designed over the past few decades, using several types of materials and processes. This textbook will primarily treat those that have become the most widely accepted by industry today.

10.4: DESIRABLE CHARACTERISTICS FOR PWB MATERIALS

PWBs are made of two very different types of materials: the substrate material, which is an insulator, and the connecting material, which is a conductor. Let's first talk about the desirable characteristics of the conductive material.

10.4.1: Desirable Characteristics for PWB Conductive Materials

First of all, the conductive material should be an excellent conductor of electricity. Second, it should be relatively inexpensive. Third, it should be easy to form into sheets that can be adhered to the substrate. Fourth, the conductive material should have a relatively low CTE, or at least a CTE closely matched to that of the substrate. Fifth, the conductive material should also be an excellent conductor of heat.

Interestingly, most of the materials that qualify for this job are elemental metals. Table 10.1 summarizes the performance of several metals commonly used in various ways in electronics. From this table, it can be seen that one of the very best candidates is copper; small surprise it is, then, that copper is so predominant in use as a conductor on PWBs.

Characteristic	Copper	Gold	Nickel	Aluminum	SAC 305
Electrical Resistivity (Ω-cm)	1.7×10^{-6}	2.2×10^{-6}	6.9×10^{-6}	2.65×10^{-6}	132×10^{-6}
Temp. Coeff. Of Electrical Resistivity Ω-cm/°C	6.8×10^{-9}	4.0×10^{-9}	-	4.3×10^{-9}	-
Thermal Conductivity (W/m·K)	4.03	3.19	.907	2.37	58
CTE (ppm/°C)	16.12	14.2	13.4	23.1	21.6
Ductility (relative comparison)	Excellent	Superb	Poor	Very Good	Very Good
Melting Point (°C)	1083	1064	1455	660	220
Elastic Modulus (GPa)	1324	77.2	2000	34.5	50
Cost (relative comparison)	Low	Very, Very High	Medium	Low	Medium-Low

Table 10.1 Characteristics of common PWB connective materials.

10.4.2: Desirable Characteristics for PWB Substrate Materials

There are many characteristics which are desirable for PWB substrate materials, and there are many materials which have been developed for this need. Some of these desirable characteristics include the following:

1. It must be an excellent electrical insulator. This means that it has a high breakdown voltage, high electrical resistivity, and low moisture absorption.
2. It must be chemically inert. It would not be good if the substrate material were to react with the atmosphere in which it was placed, or with the chemicals with which it commonly came into contact.
3. It must be physically tough (meaning it can withstand repeated impacts without breaking). PWBs must provide adequate physical support under loads of vibration and shock, for all the components placed on them.
4. It must be light. In all applications this is of some importance, but in weight-critical applications such as portable equipment and space-based equipment, this is essential.
5. It must have good thermal conductivity. The heat dissipated by the components on the PWB (especially the ICs), must be conducted away from the components so as to enable them to not overheat.
6. It must have a low CTE. As the PWB heats and cools, this causes physical expansion and contraction. The lower the CTE is, the lower the amount of expansion and contraction.
7. It must have a low dielectric constant. High-frequency signals are easily cross-coupled to other signals if there is sufficient parasitic capacitance. As discussed before (see chapter 2), the lower the dielectric constant, the lower the degree of signal cross-coupling.

The dielectric constant is also a major factor in the time it takes a signal to propagate from one part of the PWB to another. This parameter, usually termed the time-of-flight or TOF, is a function of the distance between the two points of interest, and the dielectric constant of the substrate material. The lower the dielectric constant, the shorter the TOF, and therefore the higher the frequency at which the PWB may be used.

8. It must have a low dissipation factor. As a signal propagates from one point to another, it loses some of its amplitude along the way, primarily due to dissipation losses. The lower the dissipation factor, the lower the signal dissipation losses.
9. It must be low in cost. This is always desirable, but in consumer applications, this is essential.

It is very difficult for a single material to meet all of these requirements, which is why there are so many choices when it comes to PWB substrate materials. It should be noted here that there are two components in most rigid PWB substrates: the resin and the fiber. Flexible substrates generally use only the resin. Because the resin and the fiber are so different physically, they also have very different material properties. It is best to be familiar with the relative properties of each separately; when used together, their properties generally become about equal to the average of the separate material properties. For example, if the dielectric constant of epoxy resin is 3.6, and the dielectric constant of E-glass is 6.3, the dielectric constant of FR-4 (a mixture of epoxy resin and E-glass) would be approximately 4.5. Table 10.2 summarizes some of the important properties of resin substrate materials used in modern PWBs; Table 10.3 summarizes properties of fiber substrate materials for PWBs.

Rigid PWB substrates, as mentioned in section 10.4.1, are generally made from a number of sheets of prepreg material. This material is made by taking a cloth woven of the desired fiber material, and impregnating the cloth with the desired resin by immersing the cloth in the resin. This composite of cloth and resin is then stretched flat and heated until the resin partially cures. This leaves the composite material in a dry, manageable form, which can then be laid up in sheets for the PWB cores. Because this composite material has already been impregnated with resin, it is termed pre-impregnated, or "prepreg" for short.

	Resin					
Characteristic	Epoxy	Polyimide	Cyanate Ester	PTFE	BT	Phenolic
CTE (ppm/°C)	69	50	55	99	-	-
Volume Resistivity (MΩ·cm)	10^6	10^6	10^6	10^6	10^6	10^4
Water absorption (max %)	1.0	1.5	1.0	0.20	0.80	1.3
Breakdown Voltage (kV)	35.0	40.0	40.0	20.0	30.0	15.0
Dielectric Constant (@ 1MHz)	3.6	3.2	3.1	2.0	3.1	4.8
Dissipation Factor (@ 1MHz)	0.032	0.020	0.005	0.0002	0.003	0.040
Thermal Conductivity (W/m·K)	0.19	0.18	0.20	0.19	0.20	0.37
Cost (comparative rating)	Low	Medium	Medium	Medium	Medium	Lowest
Toughness (comparative rating)	High	High	Medium	Medium	Medium	Poor
Density (comparative rating)	Medium	Low	Medium	Low	Medium	Lowest
Chemical Inertness (comparative rating)	Medium	High	-	High	-	Medium

Table 10.2 Properties of common PWB resin substrate materials.

	Fiber					
Characteristic	Paper	E-glass	S-Glass	D-Glass	Aramid (Kevlar®)	Quartz
CTE, axial (ppm/°C)	-	5.5	2.6	3.0	-5.2	0.54
CTE, transverse (ppm/°C)	-	5.5	2.6	3.0	41.4	0.54
Volume Resistivity (MΩ·cm)	10^2	10^6	10^6	10^6	10^6	10^6
Water absorption (max %)	25.0	0.0	0.0	0.0	1.0	0.0
Breakdown Voltage (kV)	5.0	30.0	40.0	40.0	30.0	40.0
Dielectric Constant (@ 1MHz)	4.8	4.8	5.4	4.8	4.2	4.0
Dissipation Factor (@ 1MHz)	0.085	0.040	0.002	0.001	0.002	0.0001
Thermal Conductivity (W/m·K)	Poor	1.0	3.0	≈3.0	0.25	1.0
Cost (comparative rating)	Lowest	Low	Medium-Low	Medium -Low	Medium	Low
Toughness (comparative rating)	Poor	Good	Good	Good	Excellent	Very Good
Density (Specific gravity)	.90	2.54	2.49	2.16	1.44	2.2
Chemical Inertness (comparative rating)	Poor	Good	Good	Good	Very Good	Good

Table 10.3 Properties of common PWB fiber substrate materials.

Sometimes the PWB needs to be made from flexible material; this is especially true for applications which call for the PWB to serve as a connector between a fixed object and a sliding or pivoting object. Examples include disk drives and moving-head printers. In these cases, the PWB is usually made from only the resin, so that it remains flexible. The most popular material for this application is polyimide, with PTFE a distant second. Figure 10.7 shows one such PWB in its application.

10.5: PWB TERMS

When it comes to PWBs, there are a few terms that may be new to us. One of these, which has already been used, is "via." In PWBs, a via is simply a connection between layers, in the z-dimension of the board. It is usually made by drilling through the layers at the points (via pads) where the connection is needed, then plating the hole with copper.

There are two other terms which are offspring of this definition: blind via, and buried via. These are shown in Figure 10.8. A blind via goes from an outer layer to an inner layer, but does not go all the way through the board. A buried via goes from an inner layer to another inner layer; it does not go to either outer layer. A multilayer PWB has three or more layers (usually at least four). A "land" is simply the physical copper going from one pad or via to another, on any of the layers.

Figure 10.7 A polyimide PWB.

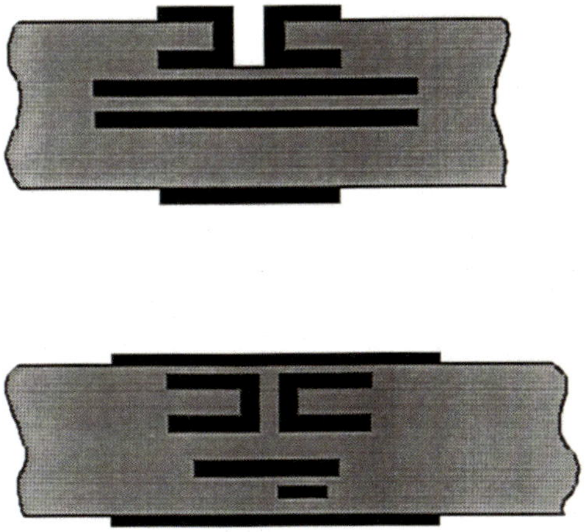

Figure 10.8 (a) A blind via, and (b) A buried via. *Reprinted from Handbook of Electronic Package Desing, p. 116, courtesy of Marcel Dekker, Inc.*

10.6: FUTURE NEEDS AND CHALLENGES

PWBs can be made in essentially any shape or size. Some have been made as small as ½" x ½", or as large as 48" x 36". Typical sizes range from 3" x 5" to 10" x 15". The thicknesses also vary significantly, with flexible PWBs coming in as thin as 2.5 mils and as thick as 15 mils (not quite so "flexible"!) Rigid PWBs range in thickness from 20 mils (considered a very thin PWB) to as thick as 250 mils, with typical thicknesses being 40–50 mils. Current materials cover this range of sizes and thicknesses quite well, and while improvements will undoubtedly continue in this area, it is currently not a major limit to future growth.

Most PWBs made still only have two layers of wiring, but this trend is changing. The number of PWBs with only one layer is still declining, and the number of PWBs with over two wiring layers is growing significantly. Typical multi-layer PWBs have from four to eight wiring layers. Future PWBs will almost certainly have more wiring layers. While this is not technically challenging, because we have been making multi-layer PWBs for many years, it is a challenge for cost. A PWB with four layers is more than three times as expensive as one of equal size with only two layers. The cost goes up significantly as the number of layers increases, and it is not a linear increase. The challenge for the future is to find a way to design and manufacture multi-layer PWBs for less cost.

The operating frequency of IC circuits continues to go up as well. This means that the PWBs the ICs go on must also be able to increase in high-frequency performance. This is a challenge both for PWB designers and for PWB materials, since PWBs capable of the very high frequencies are currently much more expensive than their lower-frequency cousins.

Another challenge for future PWBs is the pitch required between pads, particularly for ICs with periphery-lead packages. As the lead-pitch on IC packages grows finer and finer, so does the required resolution and cleanliness for the PWB manufacturing process, which in turn increases the cost of the PWBs.

A final area where future performance will need to improve is in the area of heat dissipation. ICs are dissipating more and more power per chip package, and much of this heat is passed directly from the IC chip to the PWB through the I/O leads. In turn, the PWB must be able to dissipate this heat to the surrounding air. The significance of this challenge will also continue to grow.

CHAPTER SUMMARY

In situations where size, weight, or performance are critical design factors, it is often possible that the design of an MCM or hybrid would be less expensive than the multi-module and discrete components alternative. These level 1½ devices represent a great investment of NRE costs, and are not often justifiable except at high volumes or in specialized applications.

PWBs (level 2) are ubiquitous; it is rare that an electronic product today does not have at least one of them. One and two-layer PWBs are still the most common, simply because they are much less expensive than multi-layer PWBs, but the trend continues toward more and more multi-layer PWBs because of the complexity of the devices they interconnect.

The PWB itself provides mechanical fixation for the components, electrical interconnection of the components, electrical separation of the components, and a significant path for removing the heat generated by the components. There are many characteristics which are desirable for PWB materials; these are summarized in tables 10.1, 10.2, and 10.3.

FOCUS QUESTIONS

1. What are the main advantages of a hybrid or MCM?
2. What is the difference between a hybrid and an MCM?
3. Why are hybrids and MCMs often referred to as packaging level 1½?
4. What kind of capacitors are most common on MCMs and hybrids? Why?
5. What two kinds of resistors are most common on MCMs and hybrids? Why?
6. Explain why the issue of known good die (KGD) is so significant with hybrids, and especially with MCMs.
7. The term PWB ("printed" wiring board) is somewhat misleading, because they are not printed. What term would you use instead of "printed" to better describe the process?
8. What is prepreg?
9. What is a core in a multi-layer PWB?
10. Explain why a low CTE is desirable in PWB materials.
11. Explain why a high thermal conductivity is desirable in PWB materials.
12. Explain why a high volume resistivity is desirable in PWB resin and fiber materials.
13. Explain why a low dielectric constant is desirable in PWB resin and fiber materials.
14. In what type of applications would it be important to look for a PWB made of materials with a low density?
15. Suppose a PWB is to be designed with a land that is to carry 10.0 Amps, but due to size constraints, there is not room to make the land as wide as it needs to be. What is one good alternative?
16. If a PWB is to be designed with a land that is to carry 7.3 Amps, in an ambient temperature of 40 °C, and with a temperature rise of no more than 45 °C. What would be the required land width with 2-oz copper? With 1-oz copper?

17. If a PWB is being designed with standard 1-oz copper and a maximum land width of 20 mils, what is the maximum current that the lands can carry if they are to have a rise above ambient of less than 30°C?
18. A given PWB is designed with ½-oz copper and a maximum current of 3.0 Amps. What would be the rise above ambient of a 30-mil land carrying the maximum current?
19. An uncoated PWB is to be designed for a consumer application (below 10,000 ft). What is the recommended spacing between lands for a maximum voltage difference of 250 Volts?

FOR ADDITIONAL READING

Blackwell, Glenn R., "The Electronic Packaging Handbook", Chapters 5, 7, CRC Press, Boca Raton, FL, 1999.

Harper, Charles A., "Electronic Packaging and Interconnection Handbook", Chapter 7, 8, McGraw-Hill, New York, NY, 1997.

Lau, John H., C.P. Wong, J.L. Prince, and W. Nakayama, "Electronic Packaging: Design, Materials, Process, and Reliability", Chapter 2, McGray-Hill, New York, NY 1998.

Matisoff, Bernard S., "Handbook of Electronics Packaging Design and Engineering", Chapter 4, Van Nostrand Reinhold, New York, NY, 1990.

Pecht, Michael, R. Agarwal, P. McCluskey, T. Dishongh, S. Javadpour and R. Mahajan, "Electronic Packaging: Materials and Their Properties", CRC Press, Boca Raton, FL, 1998.

Seraphim, Donald P., R.C. Lasky, and C. Li, "Principles of Electronic Packaging", Chapters 4, 10, 11, 16, 17, McGraw-Hill, New York, NY, 1989.

Tummala, Rao R. and Eugene J. Rymaszewski, "Microelectronics Packaging Handbook", Chapter 12, Van Nostrand Reinhold, New York, NY, 1989.

CHAPTER 11: PWB ASSEMBLIES (LEVEL 2)

CHAPTER OVERVIEW

This chapter covers the following concepts:

- Through-hole and surface-mount PWB assemblies
- Mixtures of through-hole and surface-mount PWB assemblies
- Processes used in making through-hole, surface-mount, and mixed PWB assemblies
- The processes of wave soldering, reflow soldering, and hand soldering

11.1: INTRODUCTIOIN

PWB assemblies consist of the PWB and all the components, (electrical, mechanical, optical, or other), that reside permanently on the PWB. Worldwide, millions of these assemblies are manufactured every day through a manufacturing process that is quite well understood and is practiced in essentially the same way wherever it is done. Many are assembled in job shops, more commonly called contract manufacturers (CMs); the remainder are assembled in facilities belonging to the company that has designed or is producing the product into which the PWB assembly will go.

The range of components that can go on PWBs is huge, and includes the following: ICs, resistors, capacitors, potentiometers, diodes, inductors, solenoids, motors, relays, fuses, breakers, optical devices, transducers, transformers, crystals, headers, sockets, switches, and wires. Also, in some applications, the IC chip is attached directly to the PWB, obviating the need for a level 1 package; this process is known as Direct Chip Attach (DCA) or Chip On Board (COB).

PWB assemblies can be categorized many ways. One of the most helpful ways, for this chapter, is to categorize them by the mounting technology of the components that are on them. These mounting technologies are through-hole-mount technology and surface-mount technology.

11.2: THROUGH-HOLE MOUNT TECHNOLOGY

Through-hole-mount technology has been with us since the invention of PWBs. In this technology, all device I/Os are in the form of a pin which goes through a hole in the PWB, and which is later soldered to the PWB. Because it is the older mounting technology, it will be covered first.

The first step in assembling a PWB with through-hole components is to populate the PWB with the components. For low volumes, this procedure is done by hand, but this is tedious work. For this reason, the process is usually performed by highly specialized robots known as pick-and-place robots. There are pick-and-place robots for axial components, radial components, DIP ICs, PGA ICs, and even odd-form components. The pick-and-place robots clench the leads of the components after inserting them into the holes, so the components do not readily fall out. After placement of components by the pick-and-place robots, components that cannot be placed by a pick-and-place robot are placed by hand. All excess lead lengths are also removed at this time.

The PWB loaded with components is now ready for wave or flow soldering. This operation is diagramed in Figure 11.1. The assembly is first heated some, then coated on the pin side with flux, to assist in the soldering process. More on soldering will come later.

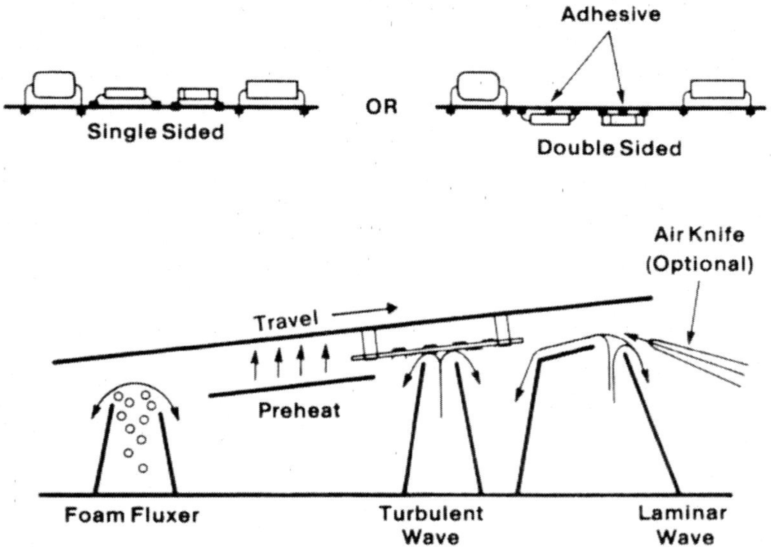

Figure 11.1 The wave soldering process. *From Tummala, Rao R. and Eugene J. Rynaszewski. 1989. Microelectronics Packaging Handbook. New York: Kluwer Academic Publishers.*

The assembly is then heated further, to about 70°C less than the temperature of the molten solder (about 250°C). This is to lessen the thermal shock when the assembly encounters the molten solder wave. The next step is to pass the assembly over the flowing wave of molten solder, upon which the solder adheres to the leads, fills the holes, and wicks up through the holes to secure the leads in the holes from the top to the bottom, which makes a very strong physical connection and a very reliable and low-resistance electrical connection. Finally, the assembly is cooled, then removed from the wave-solder machine.

The final step is to wash the flux off the assembly, which is done in what is basically an in-line dishwasher. The board is sprayed with de-ionized (DI) water, which has often been mixed with other cleansing agents. The board is then rinsed and air-dried, usually with forced air to remove the water beneath and between components.

This process has been used for many, many years to assemble PWBs with through-hole components. Although there are still many ways for the process to go wrong, with proper maintenance and oversight this process can produce PWB assemblies with excellent solder connections and very high reliability.

11.3: SURFACE-MOUNT TECHNOLOGY

Surface-mount technology (SMT) is a relatively new technology, having had its start in the early 1980s. It quickly gained popularity, in spite of the initial challenges it had. While initially used only in space or weight-critical applications, it is now used in most PWBs, to at least some degree. The advantages of surface mount technology include the following:

1. It allows components to be placed on both sides of the PWB, essentially doubling the density
2. SMT components are less expensive than their through-hole counterparts
3. SMT components are smaller and lighter than their through-hole counterparts.

With all these advantages, it is easy to see why SMT has overtaken through-hole technology in most applications.

Initially, only resistors, non-electrolytic capacitors, transistors and diodes were offered in SMT. While these components still constitute the vast majority of SMT components produced, they are no longer the only ones on the market. A good selection of potentiometers, inductors, electrolytic capacitors, fuses, relays, optical devices, switches, headers, and crystals may now be found in SMT. Figure 11.2 shows several SMT components currently available.

The first step in making a SMT PWB assembly is to deposit solder paste on the pads of the PWB where the SMT components will be placed. This is usually done by a screen-printing operation, depicted in Figure 11.3. A stainless-steel screen is made to have holes of the right shape and size in just the places where the solder paste is to be deposited. The solder paste is then forced through these openings and onto the PWB. The screen-printing of solder paste, when it works well, is an excellent high-volume

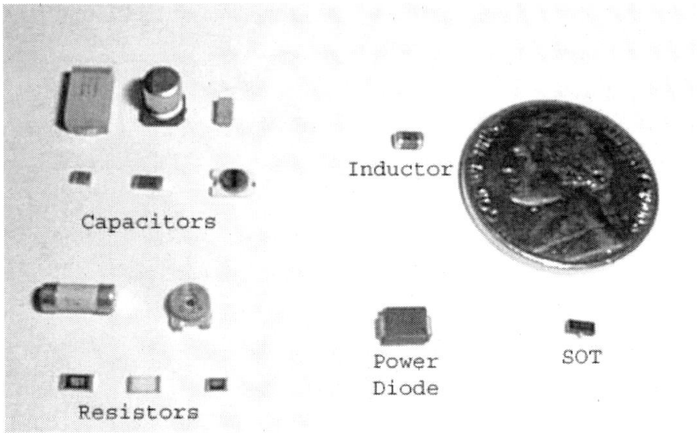

Figure 11.2 Several SMT components are currently available.

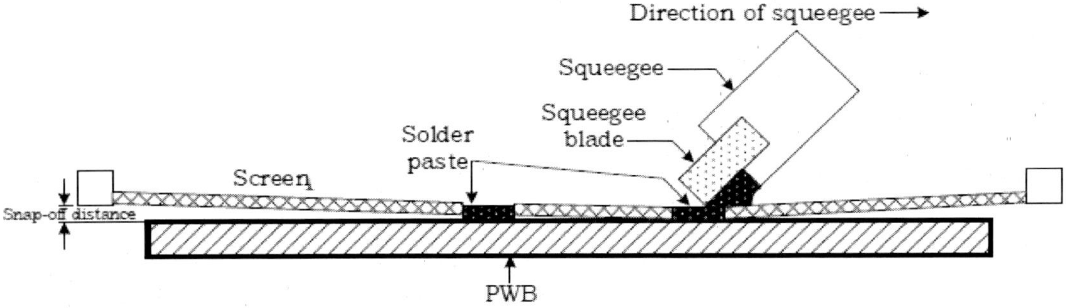

Figure 11.3 The solder paste screen printing process.

that it is the most troublesome of all the operations in SMT assembly. Studies show operation, but it should be appreciated that the quality of this operation depends on over 50 parameters, which makes it a very tricky process at times. Studies also show that the majority of test fails of PWB assemblies are due to problems that can be traced to this procedure.

The next step is to place the components onto the PWB, with the pads of the SMT components being placed directly on top of the screen-printed solder paste. This operation is usually done by high-speed pick-and-place machines, some of which can place up to twelve components per second (43,200 components/hour, or cph). More typical are the ones that place in the range of 5,000 to 10,000 cph.

After the components are placed onto the solder paste, the assembly is processed through a reflow oven. The reflow oven, much like a high-technology pizza oven, carefully brings the temperature up as the assembly passes through it. The initial heat activates the flux, cleaning the pads to be soldered. At the peak of the temperature ramp, the solder in the solder paste melts, adhering to the pads of the component and to the pads on the PWB, forming an excellent mechanical, electrical, and thermal connection. The temperature is then lowered, and the PWB assembly emerges from the reflow oven essentially complete.

In many cases, the other side of the PWB must also have components attached to it. This is done in either of two basic ways. In one way, the process described above for the first side is simply repeated for the second side. As the inverted assembly passes through the reflow oven for the second time, the previously-reflowed components, now on the bottom side of the PWB, are held in place by the surface tension of the solder. In the other way, the side which will be the bottom in the reflow oven is treated very differently. In this way, glue dots are first dispensed onto the PWB between the pads of the SMT components to be placed, as shown in Figure 11.4. Then the components are placed onto the glue on the PWB and the glue is dried (usually either by heat or by exposure to UV). With the components securely glued onto the bottom of the PWB, the PWB is then

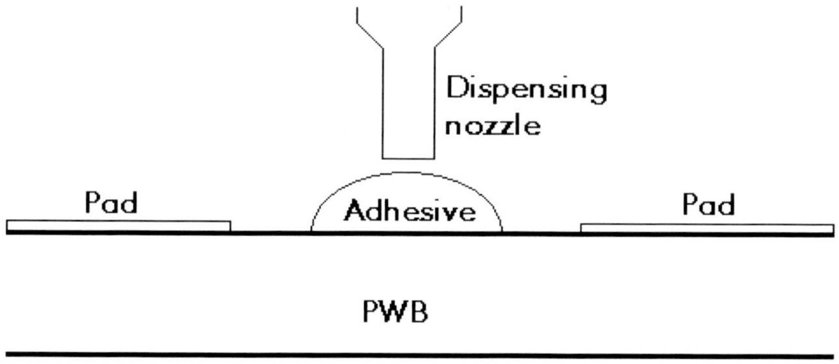

Figure 11.4 Placement of an adhesive dot for SMT mounting of bottom side components.

wave soldered. After the bottom-side components are attached in this way, the top side proceeds as described in the preceding paragraphs. The final step is to wash the board; this step was described in the previous section.

11.4: MIXED MOUNT TECHNOLOGY

In today's electronic manufacturing, very few assemblies are either completely through-hole technology or surface-mount technology. This is what we mean when we refer to mixed mount technology, and most PWB assemblies manufactured today are mixed mount technology.

There are two main methods which are currently used to assemble mixed mount technology boards. The first involves procedures which are a combination of the preceding two sections. In this method, the first step is to screen-print solder paste, place SMT components, and reflow the "top" side of the board, as described in section 11.3. Next, the assembly is flipped, and the surface which does not have any of the through-hole components (the "bottom" or "pin" side) is dotted with glue dots, the components are placed, and the glue is cured. Then the assembly is flipped again, the top side of the assembly receives its through-hole components, (the SMT components have already been placed and reflowed) and the assembly is wave soldered. In this case, the wave soldering operation solders both the top-side through-hole components and the bottom-side SMT components (held in place by glue dots).

The second method in use is known by several names, including reflow of through-hole (ROT), through-hole in solder paste (TSP), and others. This procedure involves screen printing, populating, and reflowing the bottom side first, then screen-printing the entire top side of the PWB, even where the through-hole components will be placed. Screen-printing the solder paste is followed by populating the top side with both the through-hole components and the SMT components, after which the entire assembly is again reflowed. Although there are many challenges in getting this approach to work as needed, the successes have reported have been significant. And because it saves steps in the manufacturing process, it is bound to gain in popularity in the coming years.

As with the preceding two sections, the final step is to wash the entire assembly. The final product is a level 2 PWB assembly, ready to be tested and put into service.

11.5: SOLDERING

Soldering has been around for well over two millennia; relics from before Christ have been found to have used solder. In spite of all that has been learned about soldering over the centuries, in the applications presently used in electronic manufacturing, there is still much room for improvement, especially in screen-printing and reflow of solder paste. There are three main ways in which soldering is done in electronic manufacturing: wave soldering, reflow soldering, and hand soldering. The first two have been discussed briefly in the preceding sections; more details will be given here, and the process of hand soldering will also be discussed.

11.5.1: The Process of Soldering

The process of soldering usually involves the following four components: the base metal (the metal to which the solder will attach), flux, solder, and heat. A good solder joint is strong mechanically, and is an excellent conductor of electricity and heat.

Soldering is very different from adhesive bonding, in that the solder forms an alloy at the interface with the base metal. This interface is where the constituent elements change from the base metal, then to intermetallics, then to the solder. Intermetallics are various alloys of the base metal with the solder, and are brittle and have higher resistance; it is therefore desirable to keep this layer of intermetallics as thin as possible.

One major factor in making good solder connections is to have the base metal surfaces free of contaminants and oxides. Because oxides form readily on most metals, there needs to be a way to remove these oxides. Also, oxides form much more rapidly on base metals at the elevated temperatures necessary for soldering; there needs to be a way to prevent these oxides from forming as the assembly is heated for soldering. Fluxes accomplish these tasks in two ways. First, the flux usually contains an acidic compound which removes the base metal oxides that have formed over time. Second, the flux coats the base metals and prevents the atmospheric oxygen from reacting with them as the assembly is heated. Fluxes also help lower the surface tension of the solder, which makes it easier for the solder to wet the desired surfaces. Fluxes are also the main reason PWB assemblies are washed after soldering; if the acidic fluxes are not removed, they will continue to corrode the metals on the components and the PWB.

Another major factor in making good solder connections is the thermal cycle. The ideal thermal cycle is a careful tradeoff between conflicting parameters necessary for quality. Too quick of a thermal ramp-up and decline causes undue thermal stresses on the assembly as it first expands and then contracts. Too fast of a cool-down can cause crystallization and embrittlement of the solder. Too slow of a thermal ramp-up gives additional time for oxides to form, for the intermetallic layer to thicken, and for the base metals to react with the solder and possibly contaminate it.

The most common solder in use today is a near-eutectic solder known as SAC 305, made of 96.5% tin (Sn), 3.0% silver (Ag), and 0.5% copper (Cu), and which melts at a temperature of 220°C. Eutectic means that as it is heated, it passes from a solid stage to a liquid stage without passing through a pasty stage. This is desirable, because this way the solder spends a minimal amount of time in the stage where intermetallics form, and where components can shift while the solder cools. Other near-eutectic alloys of tin, silver, copper are also used.

A high-quality solder joint is characterized by a shiny surface and a full fillet between angles on the pads and components, as shown in Figure 11.5. The most common solder defects are too much solder, insufficient solder, solder bridges, and solder balls. Additionally, solder joints may be defective if they are too porous or too grainy, if they are incompletely wetted, if they have been overheated, or if they have not been adequately cleaned.

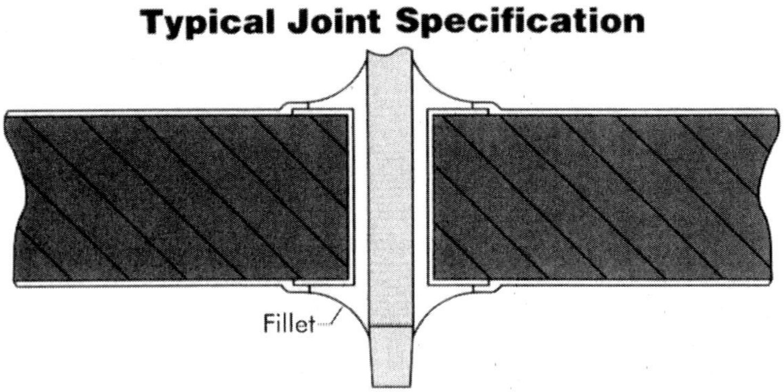

Figure 11.5 An example of a good quality solder joint. *From G. Rutter and D. West. June, 1997. Electronics Production & Packaging. Used by permission.*

11.5.2: Wave Soldering

The process of wave soldering was described previously in section 11.2. The entire process of passing a PWB assembly through a wave-solder machine is automated by a conveyor system which holds the sides of the assembly as it passes through. The flux is deposited onto the bottom surface of the PWB by either spraying it on or by passing the PWB assembly over a foam flux wave. An air knife is sometimes used next to remove excess flux. The assembly is then heated, both to prepare it for entering the solder wave, and to activate the flux so it can quickly remove any base metal oxides.

The next step is for the assembly to pass over the solder wave. Many modern wave-solder machines have two waves, a turbulent wave and a smooth or laminar wave. The purpose of the turbulent wave is to move the solder into areas that are sometimes blocked or shadowed by bottom-side components. The laminar wave is second, so as to smooth the amount of solder deposited on all the exposed base metals. The final step is the cooling-down stage.

11.5.3: Reflow Soldering

The process of reflow soldering was described previously in section 11.3. The quality of the reflow process depends on many parameters, including the following:

1. the temperature cycle
2. the accuracy of the screen printing process
3. the accuracy of the pick-and-place process
4. the process time of the assembly between screen-printing of the solder paste and its entry into the reflow oven

5. the type of solder resist used
6. the type of solder paste used.

Temperature Cycle

The challenge of the ideal temperature cycle was discussed in section 11.3. In most cases, this parameter is well controlled and is not a significant problem.

Accuracy of Screen Printing Process

It is impossible to correctly reflow solder paste onto the desired pads and components if the paste has been incorrectly printed onto the PWB. Careful inspection after screen printing is warranted to verify this accuracy before proceeding further. Most high-volume producers use optical inspection for this.

Accuracy of Pick-and-Place Process

As with the previous parameter, it is impossible to correctly reflow the components if they have not been correctly placed onto the PWB. Again, optical inspection is very often used to avoid the challenge of costly rework.

Time Between Screen-Printing and Reflow

Once the solder paste has been printed onto the board, it begins to undergo undesirable changes affecting its reflow performance. These changes have mostly to do with the loss of volatiles in the flux and binder of the solder paste. Modern solder paste formulations can wait up to about eight hours before being reflowed.

Type of Solder Resist Used

During the process of reflow soldering, occasionally a small ball of solder will shoot off from the paste as it melts, instead of melting with the rest of the paste on a surface. This little ball of solder, once it cools, is a problem waiting to happen. If it does not cause a short immediately, it can often do so later when it is dislodged and rolls to another place. Studies show that the type of solder resist used has much to do with the formation of solder balls. The solder resist also has a major effect on one of the other two main types of reflow solder defects: shorts.

Type of Solder Paste Used

Solder paste is a mixture of very small solder spheres, flux, binder, and rheological agents. Solder pastes come in many formulations. Several fluxes are also available, from highly active (very acidic) to mild, even to no-clean fluxes. Fluxes are also available in water-based and solvent-based varieties.

Several solder alloys are available, from the eutectic Sn/Pb (still used in many military applications) to alloys containing silver, indium, and bismuth. Most common today are alloys of tin, silver and copper, with SAC 305 being one of the most popular. Several sizes of solder spheres are also available; the smaller the spheres, the more expensive the solder paste, but very small spheres are essential for fine-pitch peripheral-lead IC packages.

The binders and rheological agents (chemicals that determine the flow, printing, and slump characteristics of the paste) are essential to the success of the paste. The binder holds the spheres and the flux together, to keep them from separating out, and also provides tackiness to hold the components in place after pick-and-place. The rheological agents allow the paste to squeeze nicely through the screen openings during screen printing, and to stand up straight (no slumping allowed!) on top of the desired surfaces until reflow.

11.5.4: Hand Soldering

The same factors that determine the quality of solder connections in wave soldering and reflow soldering also determine the quality of hand-soldered connections. The base metal surfaces must be clean and free from oxides. The tip of the soldering iron should also be clean. And the temperature of the iron should not be excessive; somewhere between 290°C and 315°C (550°F to 600°F) usually works best.

Figure 11.6 depicts the correct sequence for hand soldering. First, a bit of solder is melted onto the tip of the iron to facilitate heat transfer between the tip of the soldering iron and the surfaces to be soldered. Next, the melted solder on the iron is brought into contact with both the pin and the pad to be soldered. After a few seconds to allow the surfaces to heat up, the wire solder is touched to one of the surfaces to be soldered; it should not be touched to the solder already on the tip of the iron. The reason for this is that the wire solder has a thin core filled with flux, and this flux is needed on the surface of the base metals to be soldered. If the wire solder is touched to the solder already melted on the tip of the iron, the flux is burned away before it touches the base metals. If done correctly, you will first notice a thin, watery fluid (a very small amount) coat the base metals, after which the solder melts and wets the base metal surfaces very nicely. After the solder has wetted all the necessary surfaces and sufficient solder has been deposited in the connection, the solder is removed, then the iron is removed, and then the joint is allowed to cool naturally. Do not blow on the joint, as this will cause the solder to cool too rapidly, leading to crystallization and embrittlement.

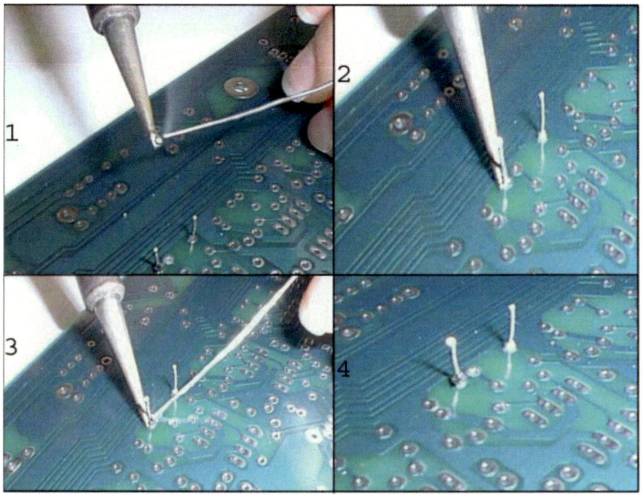

Figure 11.6 The correct sequence for hand soldering. 1) Apply a bit of solder to the iron to promote heat conduction fronm the iron tip to the joint to be soldered. 2) Apply the soldered iron to the joint. 3) Apply the solder from the other side of the joint. 4) After first removing the solder, then the iron, allow the joint to cool slowly.

Good quality hand-soldered connections take practice. Do not expect to be able to do it right on the first try. Practice on a scrap board with scrap components until your joints look like those on professionally-made PWB assemblies.

11.6: FUTURE CHALLENGES AND NEEDS

Although industry has been making solder connections by the millions for decades, it is still one of the main sources for failure in PWB assemblies, simply because there are so many of them. Some of today's soldering challenges and needs include the following:

1. Better no-clean solder paste formulations. A paste with flux that does not need to be cleaned off after reflow would save another major processing step. Present formulations use a mildly active flux, which means it requires very clean surfaces or it won't work adequately. Here is another engineering tradeoff: the more active the flux, the better the process can handle oxides on the surfaces of the base metals; but active fluxes must be thoroughly washed off, or they will impact the future reliability of the PWB assembly.
2. Better rheological agents. Present solder pastes tend to slump after being printed, and the degree of slump depends on humidity and temperature. If the paste would stay just as it was printed and not slump, the humidity and temperature of the facility would not need to be as carefully controlled.
3. Finer solder spheres for a lower price. Solder pastes are not cheap, starting at about $50/lb, and going as high as $130/lb for pastes with very fine spheres.

Fine-pitch IC packages are here to stay; solder pastes for these fine-pitch leads must improve and come down in cost.

4. More data on currently-used no-lead solder pastes. These solder pastes, required for ROHS compliance (reduction of hazardous substances), are still not as familiar as the tin/lead solder they have replaced, so much study needs to be done.

CHAPTER SUMMARY

PWB assemblies are composed of the PWB along with all the components it mechanically holds and electrically connects. Components are currently separated into two large groups: through-hole and surface-mount. Most PWBs have a mixture of both through-hole and surface-mount components. Although this is not ideal, it is the case simply because most components are not completely available in both types of packages.

Through-hole devices are first inserted, then soldered to the PWB using the wave-solder process. Surface-mount devices first have solder paste screen-printed onto their pads, then they are placed onto the PWB, and then the assembly goes through the reflow-solder process. Screen-printing solder paste onto surface-mount technology PWBs is a very demanding and trouble-prone process, and great care must be taken to perform this procedure correctly every time.

Surface-mount technology has become increasingly popular because the components are significantly smaller than their through-hole counterparts. Most surface-mount devices are also less expensive than their through-hole counterparts, and this will continue. For size-critical applications (e.g., portable products), surface-mount technology is the only possible way to make such small units.

FOCUS QUESTIONS

1. What is DCA (direct chip attach)?
2. What are some of the advantages of surface-mount devices, compared to through-hold devices?
3. What is the role of a pick-and-place robot in making PWB assemblies?
4. PWB assemblies are often thoroughly washed with water, after they have been assembled and soldered. How is this practical, considering that water on PWB assemblies is generally a very bad thing?
5. Surface-mount PWB assemblies have long been more expensive and difficult to make than through-hole PWB assemblies. Then why is surface-mount technology becoming increasingly popular?
6. What is the role of flux in the soldering process?

7. What is the melting point of eutectic tin-lead solder?
8. What does eutectic mean when referring to soldering?
9. What mixture of tin and lead gives the most eutectic combination?
10. What are some of the parameters that affect the reflow solder process?
11. What are some of the characteristics of a good quality solder joint?
12. What are some of the improvements needed in surface-mount assemblies for the future?

FOR ADDITIONAL READING

Harper, Charles A., "Electronic Packaging and Interconnection Handbook", Chapters 5, 8, 9, McGraw-Hill, New York, NY, 1997.
Pecht, Michael, "Handbook of Electronic Package Design", Chapter 4, Marcel Dekker, New York, NY, 1991.
Seraphim, Donald P., R.C. Lasky, and C. Li, "Principles of Electronic Packaging", Chapters 4, 16,17, 19 McGraw-Hill, New York, NY, 1989.
Tummala, Rao R. and Eugene J. Rymaszewski, "Microelectronics Packaging Handbook", Chapters 11, 12, Van Nostrand Reinhold, New York, NY, 1989.

CHAPTER 12: MOTHERBOARDS AND BACKPLANES

CHAPTER OVERVIEW

This chapter covers the following concepts:

- Motherboards and backplanes make up packaging level 3
- How most motherboards and backplanes are made
- How motherboards and backplanes are similar and different

12.1: MOTHERBOARDS AND BACKPLANES (LEVEL 3)

Level 3 of electronic packaging is the level that interconnects multiple PWB assemblies. This level of packaging is primarily made up motherboards and backplanes. Both motherboards and backplanes are made by the same process by which PWBs are made, usually out of the same material (FR4). A motherboard is distinguished by the fact that it also has level 1 devices on it, in addition to level 2 assemblies. The motherboard of today's personal computer is a classic example; it has connectors for the extra boards that may added (display driver, drive controller, sound card, etc.); it has the microprocessor, glue logic, BIOS, and other modules, and it has connectors for the SIMMs (single in-line memory modules) (see Figure 12.1). A backplane is distinguished by the fact that it generally just has connectors for PWB assemblies; it generally does not contain any level 1 devices (see Figure 12.2).

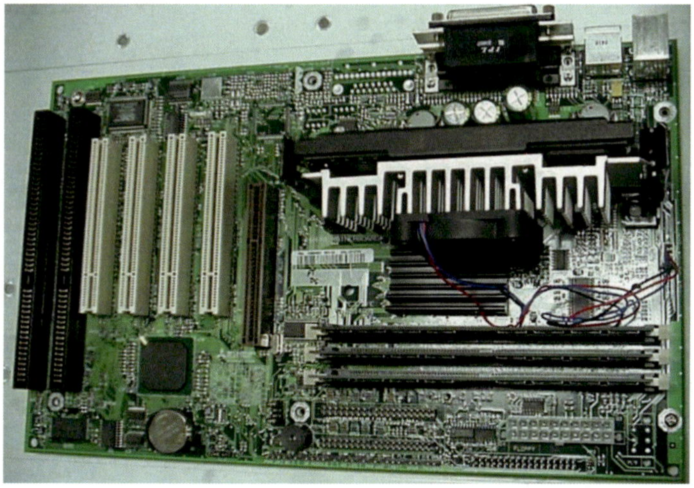

Figure 12.1 A typical motherboard. Note the many on-board components: a microprocessor with a large heat sink, a battery, various ICs, off-board connectors, SIMMS, capacitors, chip resistors and capacitors, and card connectors.

Figure 12.2 A typical backplane. Note the dominant presence of card connectors. There are very few other components; only resistors and capacitors were used on this backplane.

Connectors are an essential part of a motherboard or backplane; these were covered earlier in chapter 2. Another thing which typically characterizes a motherboard or backplane is the number of layers; while PWBs generally range from 2 to 8 layers, motherboards and backplanes typically range from 6 to 30 layers.

One of the current issues in motherboards and backplanes is that of signal integrity, especially at the frequencies of today's computers. The impedance must be predictable and carefully controlled; the crosstalk must be reduced as much as possible. The higher the frequencies on the motherboard or backplane, the bigger the design and manufacturing challenge. To address these issues, designers turn to materials with lower dielectric constant (such as Teflon®), differential signals (which take up more real estate and must be carefully laid out), impedance matching (a tricky thing to design and even trickier to consistently manufacture well), and other techniques.

Another big issue at this level is cost. The cost of motherboards and backplanes is driven by the high NRE costs of the design phase, the high costs of making these boards with so many layers and such tight requirements, and the relatively high cost of the connectors for the level two boards that they interconnect. Additionally, the materials used must be of high performance; this makes it difficult to bring those costs down as well.

CHAPTER SUMMARY

Motherboards are usually made of the same materials, and by the same processes, as PWBs, but differ in that they have other PWBs attached to them.

Motherboards also usually have some components soldered on them; in personal computers, these components include the microprocessor and several support modules.

Backplanes are similar to motherboards, except that they normally do not have components soldered to them. Backplanes are used to interconnect multiple board connectors, and are very common in rack assemblies.

FOCUS QUESTIONS

1. How is a motherboard similar to a PWB? How is it different?
2. How is a motherboard similar to a backplane? How is it different?

FOR ADDITIONAL READING

Matisoff, Bernard S., "Handbook of Electronics Packaging Design and Engineering", Chapter 8, Van Nostrand Reinhold, New York, NY, 1990.

Pecht, Michael, "Handbook of Electronic Package Design", Chapter 4, Marcel Dekker, New York, NY, 1991.

Pecht, Michael, R. Agarwal, P. McCluskey, T. Dishongh, S. Javadpour and R. Mahajan, "Electronic Packaging: Materials and Their Properties", CRC Press, Boca Raton, FL, 1998.

Seraphim, Donald P., R.C. Lasky, and C. Li, "Principles of Electronic Packaging", Chapters 16, 17, McGraw-Hill, New York, NY, 1989.

Tummala, Rao R. and Eugene J. Rymaszewski, "Microelectronics Packaging Handbook", Chapters 11, 12, Van Nostrand Reinhold, New York, NY, 1989.

GLOSSARY

ablate: to burn away with heat, typically with a laser. The resistive ink coated onto the mandrel of carbon and metal film resistors is ablated in a helical pattern, meaning that the resistive ink is literally burned away with the laser.

age drift: the degree to which an electronic component varies from its original value as a function of the passage of time. Using the figures cited in Table 3.1, a 1000-Ohm carbon comp resistor could vary by 50 to 100 Ohms in 1000 hours of operation.

amorphous: without large areas of crystalline structure. Solid materials may be classified as crystalline, polycrystalline, or amorphous. Amorphous materials tend to be less dense, less brittle, and less efficient than crystalline materials.

anisotropic: having different properties in different axes. In the manufacturing of ICs, it is highly desirable that processes take place in a preferential direction; thus an anisotropic process is very useful. Examples include reactive ion etching and ion implantation, which occur primarily in the downward direction.

annealing: the process of raising the temperature of a particular product for the purpose of allowing the intermolecular and interatomic bonds to reform, thus imparting to the material improved properties. Silicon wafers are annealed after ion implantation, because the ion implantation process tends to dislodge many atoms from their ideal locations, and even disrupt some interatomic bonds.

axial: having leads along the axis of the body of the component. Axial components are discrete components; examples include carbon composition resistors, carbon and metal film resistors, wirewound resistors, non-SMT inductors, non-SMT diodes, and many types of capacitors. Axial components are also *through-hole* components.

backplanes: Level 3 of the packaging hierarchy, consisting primarily of board connectors and a printed wiring board interconnecting the connectors. Backplanes are typically mounted into racks, the next packaging level.

ball grid array (BGA): a surface-mount type of package for an IC chip in which the electrical connections are in the form of small solder balls in the pattern of an X-Y grid. One of the biggest advantages of the ball grid array package is the very large number of I/Os available for a given pitch and package dimensions.

bare die: sometimes also referred to as naked die, this refers to the practice of selling the unpackaged IC chip to circuit board assemblers or electronic product manufacturers. Bare die are most often used in hybrid and multi-chip modules, and chip-on-board applications.

base metal: the metal beneath the upper layers. In the case of oxidation, the oxide forms on the surface of the base metal. In the case of printed wiring boards, the base metal is copper, onto which is sometimes plated nickel, gold, tin, or other metals depending on the application.

BIOS: Basic Input/Output System. The basic information a computer needs to begin operation after a cold boot is contained in the BIOS, and includes such basic functions as keyboard reads, disk I/O, and screen I/O. This is usually contained in non-volatile semiconductor memory.

blind via: an interlayer connection in a printed wiring board, going from one of the outer layers to an inner layer, but not going all the way through the board. This type of via allows for higher wiring density than ordinary through-hole vias, but is significantly more expensive. (see **buried via**)

boule: A cylindrical, monocrystalline mass of material which is the final product of crystal growth using the Czochralski method. These can vary from 25 mm (1 inch) in diameter for specialized crystals, to 300 cm (12 inches) in diameter for silicon, and from 15 cm long to over 100 cm.

bumped: a chip (die) which has very small solder balls already attached to its I/O pads is said to be *bumped*. This is used only for attachment to array-type I/Os (as in BGAs, PGAs, or DCA), as opposed to peripheral leads.

buried via: an interlayer connection in a printed wiring board, going from one of the inner layers to another inner layer, but not going all the way through the board. This type of via allows for higher wiring density than ordinary through-hole vias, but is significantly more expensive. (see **blind via**)

burn-in: the practice of confining an electronic device or assembly in an elevated temperature environment for an extended period of time for the purpose of causing failure in those devices or assemblies that are poorly manufactured and would therefore fail early if put into the field. This practice can significantly increase the field reliability of electronic devices and assemblies. (See also **elevated voltage stress screening**)

capacitance: the property by which electrostatic charge is stored. Any time two conductors are separated by an insulator, capacitance exists. The capacitance is proportional to the surface area of the conductors and the dielectric constant of the insulating material, and is inversely proportional to the distance between the conductors.

carbon composition resistors: resistors which are made of a composite of powdered carbon and a powdered insulative material such as ceramic, mixed with a binder. Carbon composition resistors have been around for over 70 years; they are inexpensive and rugged, but have relatively poor performance in tolerance, temperature drift, and age drift as compared to other resistor types.

carbon film resistor: a resistor made by depositing a film or ink of carbon resistive material on a ceramic mandrel. Because they are as inexpensive as carbon composition resistors but have much better performance in tolerance, temperature drift, and age drift, they have largely displaced carbon composition resistors. (See also **metal film resistor**)

ceramic substrate potentiometer: a three-terminal variable resistor made with a metal or ceramic film of resistance material on an alumina (ceramic) substrate. This type of potentiometer is mid-range in terms of performance (number of adjustment cycles) and cost.

chip resistors: discrete resistors in the form of a small rectangle, ranging in size from 200 x 120 mils (2012 package type) to as small as 10 x 5 mils (01005 package type). They are made with a thin film or ink of resistive material on a ceramic substrate.

clean rooms: the specially built rooms in which integrated circuits are manufactured, consisting of high-efficiency particulate air (HEPA) filters and a laminar downward airflow pattern. Extreme care must be taken in manufacturing ICs not to contaminate the circuits with particulates in the normal air we breathe, so these are filtered out.

Clean rooms are rated in classes, where class 10 refers to a clean room in which there are no more than 10 particulates per cubic foot of a size greater than 50 nm. Clean rooms are extremely expensive (over $12,000/square foot).

clench: the process of bending a through-hole lead after it has been inserted into the board, for the purpose of keeping the component in place until soldering has occurred. This is usually done by inserting the lead through the hole and past an offset anvil under the board.

COB: Chip-On-Board. A process very similar to DCA, in which an IC chip is attached directly to an FR4 or other organic substrate, without a chip package. This is still somewhat problematic for organic substrates, since the CTEs of silicon and organic substrates differ substantially. An elastomeric material is almost always used between the IC chip and the substrate to mitigate this problem.

collimated: having the rays of light all traveling in parallel. A flashlight is an example of a light source which is non-collimated, as can be seen when the light is shined at an object: the circle of light is much greater than the diameter of the flashlight lens. A laser is an example of a light source which is highly (though not perfectly) collimated; when a laser is shined on an object, the circle of laser light is only slightly larger than the diameter of the laser lens. Collimated light is highly desirable in photolithography, as it permits higher resolution than non-collimated light.

column IV elements: elements in Column IV of the periodical table of elements; these include carbon (element #6), silicon (element #14), germanium (element #32), tin (element #50) and lead (element #82). The significance of these elements lies primarily in the first three (carbon, silicon and germanium), since they are the most common elemental semiconductor materials.

column IV materials: see **column IV elements**.

condensers: simply an early name for capacitors; now largely unused except in legacy applications.

contact: (a) the point in switches or relays where electrical connection is actually made. The size of contacts in switches and relays must be proportional to the current rating. (b) the first step in the three-step process of mating the two halves of a connector. The second and third steps are *swage* and *swipe*. Contact is the step in which the two electrically conducting surfaces first come into physical contact.

contract manufacturers: companies which produce printed circuit board assemblies and even finished products for other companies. Many companies which manufacture electronic products combine both in-house manufacturing capacity and contract manufacturers.

contact resistance: the (usually) small amount of resistance that exists when two conductors intentionally touch each other to make a contact or connection. This resistance is dramatically affected (increased) by factors such as dirt, oxidation, corrosion, arcing, or contact wear.

controlled-collapse chip connect (C4): a method of electrically and mechanically connecting an IC chip to its package. The chip contains an array of very tiny solder balls; the chip is turned face-down (which is why this method is also sometimes called the "flip-chip" method) and positioned over the package, upon which the package and chip are heated until the solder balls collapse and form the electrical and mechanical bonds. This method was pioneered by IBM in the 1960s, and is now popular with nearly all manufacturers.

crosstalk: the tendency of one signal to couple into a nearby signal, by inductive or capacitive coupling. This is undesirable, and is a form of noise. It is most often noticed in phone conversations when one can faintly hear some other conversation going on. It is particularly a problem near high-impedance nodes, or where a strong signal passes by a much weaker signal.

CTE: Coefficient of Thermal Expansion. Almost all materials expand when heated, then shrink when cooled back down. The coefficient of this expansion is usually expressed in ppm/°C. For example, high-purity alumina has a CTE of 6.7 ppm/°C, which means it expands 670 ppm (.67%) for each 100°C.

current: the flow of electrons, measured in Amps. One Amp is equal to 1 Coulomb/sec, where 1 Coulomb is equal to 6.242×10^{18} electrons. For example, if 6,242,000,000 electrons flow past a given point in 1 second, this is equivalent to 1 nCoulomb/sec, or 1 nAmp.

current rating: How much current a particular device is rated to handle. Current flow always generates heat in ordinary (non-superconducting) materials, and the larger the current, the greater the amount of heat generated. Thus the larger the current rating of a device, the larger it tends to be.

current surges: Depending on the load, the current can sometimes peak at values much higher than the normal amount of current. These peak values (current surges) can cause problems in electronic devices and products.

Czochralski method: the most common method used to grow large boules of pure crystalline silicon. Also used to grow some other semiconductor crystalline boules. It involves contacting a container of molten material with a seed crystal, then slowly rotating and removing the seed crystal while the molten material solidifies onto the seed crystal and the crystal grows. Using the Czochralski method, a nearly pure silicon boule weighing about 30 kilos can be grown in only a few hours.

DCA: Direct Chip Attach. The process by which a *bumped* chip is directly attached to a board, without a chip package. This is not too uncommon for alumina substrates, where the CTEs of the alumina and the silicon are relatively close, and it is increasingly common though still somewhat problematic, for FR4 substrates, where the CTEs of the FR4 and the silicon are substantially different.

de-facto standard: a standard which has come about not by decree of a standards-setting body, but simply by wide market acceptance. The 3½" diameter of most hard disks is an example of a de-facto standard, as is the qwerty-style keyboard common on most computers.

dendrite: derived from a Greek word for tree (*dendro*), this refers to things which are tree-like in structure. In this text, this term refers specifically to the crystalline growth products of electrolytes and other contaminating materials. Most dendrites are made of ionic materials and are therefore conductive.

dielectric: an insulating material. Most dielectrics have a resistivity in the range of 10^{10} Ohm-cm to 10^{18} Ohm-cm. Families of dielectrics include plastics, rubbers, ceramics, and glasses.

dielectric absorption: some of the energy stored in capacitors is actually absorbed within the dielectric material of the capacitor, and is therefore not readily available when the capacitor is discharged. A capacitor with high dielectric absorption is characterized by having a small amount of voltage remaining after having been completely discharged.

dielectric constant: the degree to which a particular material enhances the storage of electric charge, as in a capacitor. The dielectric constant is relative to the absolute permittivity of free space, which by definition has a dielectric constant of 1.0. All other common materials have a dielectric constant greater than 1.0.

dielectric strength: the ability of a dielectric material to withstand an electric field without breaking down. When the dielectric strength of a material is exceeded, it characteristically causes a short-circuit while the voltage field discharges through the dielectric. The classic example is when the dielectric strength of air is exceeded by building clouds, and we see the breakdown in the form of lightning.

diffusion: the permeation of a given material into another. Gases can diffuse into silicon, especially when silicon is at an elevated temperature. Diffusion is a common way to create the necessary *p*-doped and *n*-doped regions in integrated circuit transistors.

diode outline: the physical dimensions of a diode package type. These are preceded by the prefix DO-, and the various diode outlines are standardized by JEDEC. (See also **transistor outline**)

dissipation factor: the ratio of the reactance to the resistance (X_C/R_C) of a capacitor. It is the property of a capacitor which causes it to dissipate a small amount of energy. In an ideal capacitor, the resistance of the dielectric is infinite, so the dissipation factor is 0.0, meaning that none of the energy is lost to heat dissipation. In all practical capacitors, the resistance of the dielectric is finite, so a small amount of current flows through the dielectric of the capacitor. Dissipation factors for practical capacitors range from 0.8 for phenolics to 0.0002 for teflon.

dopant: to make transistors, there must be *p*-doped and *n*-doped regions in the silicon. Diffusion with a material which has fewer electrons in its valence shell than silicon (such as boron or aluminum) will produce a *p*-doped region; diffusion with a material which has more electrons in its valence shell than silicon (such as phosphorous or arsenic) will produce an *n*-doped region. Materials used for this diffusion process, called “doping”, are dopants.

doping: see **dopant** above.

drift: Many parameters which describe electronic devices or products are not stable with respect to other variables, and therefore exhibit drift. For example, temperature drift of frequency means that the frequency changes as a function of temperature. The most common parameters which affect electronic devices or products are temperature, voltage, aging, and humidity.

dry (plasma) etching: In the process of making integrated circuits, it is often necessary to remove one material from designated areas of the chips. Dry or plasma etching does this by exposing the unprotected areas of the chip to a highly energetic and reactive plasma, which reacts with the unprotected areas of the chip and etches them away. Although dry etching is more expensive than wet etching, it has the major advantage of being an anisotropic process, which permits much better resolution than wet etching, which is an isotropic process. (See also **wet etching**)

ductility: the property of a material which describes its ability to be drawn out into a fine wire. Gold has extremely high ductility, while tungsten has low ductility.

electrical overstress: this is to subject an electronic device or assembly to a voltage much greater than that for which it was designed and rated. Two common ways in which electronic devices or assemblies are electrically overstressed are electrostatic discharge (ESD) and fault voltages generated by failures in circuits that supply the operating voltages. Electrical overstress can significantly reduce the reliability of a device or assembly.

electrolyte: a conductive fluid. In this text, there are good electrolytes, such as those which are found in electrolytic capacitors, and bad electrolytes, such as those that remain as impurities on the surface of PWBs or ICs. Human spittle and finger oils are classic electrolytic contaminants.

electromagnetic compatibility: if an electronic product neither emits undue amounts of electromagnetic radiation nor fails when subjected to the same, it possesses electromagnetic compatibility. Most countries have a government agency charged with overseeing this issue, which permits electronic products to operate to known standards and provides strict limits to the interference they are allowed to produce and specifies the levels which they must be able to tolerate.

electromagnetic radiation (EMR): electromagnetism, one of the 4 fundamental forces of nature, can propagate waves of energy, which in a vacuum travel at the speed of about 300,000 km/sec. Light is the most familiar form of EMR; other forms include infrared, microwaves, and radio waves (at frequencies lower than light) and ultraviolet, x-rays, gamma rays, and cosmic rays (at frequencies higher than light).

electron and hole mobility: how freely electrons and holes (the counterpart to an electron) can move about in a given material. This mobility is a key property in determining the frequency capability of a semiconductor material. For example, the electron and hole mobility of gallium-arsenide (GaAs) is much greater than that of silicon, which makes GaAs capable of much higher frequencies than silicon.

electrostatic discharge: the brief and potentially violent release of energy stored in an electrostatic field, as in a capacitor. Intentional electrostatic discharges include the flash of a camera or the jolt of electricity sent into the paddles of a defibrillator. Unintentional electrostatic discharges include lightning and the sometimes painful shocks we receive when touching a grounded object (such as a light switch) after building up a static charge by walking on a carpet. Voltages as low as 50 Volts can damage electronic devices; most electrostatic discharges occur at voltages much greater than this (in the range of 400-15,000 Volts).

elevated temperature stress screening: see **burn-in**.

elevated voltage stress screening: the practice of operating an electronic device or assembly at an elevated voltage for an extended period of time for the purpose of causing failure in those devices or assemblies that are poorly manufactured and would therefore fail early if put into the field. This practice can significantly increase the field reliability of electronic devices and assemblies. (See also **burn-in**)

eutectic: a metal alloy with a specific melting point which is lower than the melting point of any of its constituent metals. The classic example is the 63/37 mix of Sn/Pb commonly used for decades in electronic soldering applications; its melting point is 183°C, while that of pure tin (Sn) is 232 °C, and that of pure lead (Pb) is 327°C. Tin/lead is no longer used in most electronic manufacturing, due to the efforts to remove lead from electronics assemblies.

evaporation: a method of depositing metals on the surface of an integrated circuit during its manufacturing process. Evaporation requires an elevated temperature and a low pressure (vacuum) environment. Evaporation is commonly used to deposit aluminum.

explicit standards: standards which are formally stated (written) and agreed to by relevant parties. Also known as *de juro* standards. Examples include the RS-232 interface standard, the Internet's TCP-IP, and the 120 Volts @ 60 Hz power delivered in the United States. Explicit standards are contrasted to *de facto* standards, which are standards that came to exist simply by informal, widespread adoption. An example of a de facto standard would the use of USB for a computer interface – it is commonly done, but is not an explicit requirement.

fiber substrate materials: most printed wiring boards are made from a composite of both a polymer-type of material (such as epoxy or polyimide) and a fiber (such as fiberglass or aramid). The latter constitute the fiber substrate materials.

fibrillation: the arrhythmic (random) quivering of the heart muscle. Coronary fibrillation prevents the heart from actually moving the blood, and will result in the death of the victim unless it is interrupted and converted back to rhythmic coronary contractions, usually by some external agent (defibrillation).

flow soldering: see **wave soldering**.

flux: a substance used to prevent or remove oxidation from metal surfaces. A small layer of oxide forms quickly on most metals, and will prevent a good solder joint from forming; flux aids in removing this layer and providing an excellent quality solder joint. To function well, a flux must be somewhat corrosive, which means they must be thoroughly removed after the soldering operation is complete, to prevent further corrosion.

frequency range: the low to high limits specified for the frequency of operation of an electronic product. For example, the frequency range of an FM radio receiver should be 88.1 MHz to 108.1 MHz.

fundamental physical limit: the smallest or lowest a given parameter can reach before the presently understood laws of physics prevent further reduction. The last 40+ years of Moore's Law have come about by reducing the physical dimensions of the IC transistor, but eventually we will reach a point where physics will not allow an IC transistor to properly function. This is its fundamental physical limit. We are presently very close to reaching this limit for today's MOSFET transistors, but thankfully, other devices are being developed.

hermetic: sealed against the environment. An hermetic package does not allow the ingress of moisture, dust, or other gases. Hermetic packages are the ideal in electronics, but are always more expensive than non-hermetic (usually meaning plastic) packages.

hermeticity: the degree to which a package is air and water tight. After manufacturing an IC chip, it must be protected from elements that would shorten its life, which include water vapor, liquids, or corrosive gases. Package types which are considered to be hermetic are metal and ceramic packages; plastic packages are not hermetic packages.

hot air solder leveling (HASL): the process of coating exposed metal surfaces of a printed wiring board (PWB) with solder by immersing the PWB in a pot of molten solder, and while removing the PWB, blasting hot air at it to remove the excess solder. This process has been used for many years and is reliable and inexpensive, but it deposits a layer of solder which is not highly consistent in its thickness. Where thickness dimensions are critical, it cannot be used. (See also **organic solderability preservative**)

hybrid module: a module having more than one type of device on the same substrate. Hybrids may consist of die, chip resistors or capacitors, and film resistors, as well as other SMT devices.

input/output (I/O): in an IC, the I/Os are the pins; on a printed wiring board the I/Os are the board edge contacts. Getting a large number of I/Os in a small amount of space is an industry focus in electronics, and the need for improvement in this area will continue.

insertion force: mating a chip with a socket or a plug with a receptacle requires a mechanical force to overcome the swage action of the connection. This force is known as the insertion force, and is usually measured in grams/pin or ounces/pin. In sockets with over 100 pins, the total force necessary to complete the insertion is not trivial.

integrated circuit resistors: most electronic circuits include resistors, and this is no different for integrated circuits. But in this case, the resistors are built on the same piece of silicon as the transistors and the interconnections for them. Integrated circuit resistors are the most common type of resistor, in terms of raw numbers, but most people will never see one because of their incredibly small size and the fact that one cannot see inside most IC packages.

intermetallics: the alloys of different proportions that form at the interface between two metals. In electronics, the main problems with intermetallics are that they are more brittle than the base metals from which they are formed, and they have a higher electrical resistance, thus contributing to the overall resistance of an electrical path.

ionization: the removal or addition of an electron to an atom which is normally electrically neutral. A voltage field can readily ionize neutral atoms, which increases the conductivity of the subject material very readily. The ionization of a dielectric will destroy its insulating properties.

isotropic: the opposite of *anisotropic*. An isotropic process proceeds equally in all directions, without a preferential axis. This is undesirable in IC manufacturing because it limits the type of structures that can be created, and worsens the resolution of the process.

JEDEC: formerly the Joint Electron Device Engineering Council, now just JEDEC, is the semiconductor engineering standardization body of the Electronic Industries Alliance (EIA), a trade association that represents all areas of the electronics industry (from the JEDEC home page, at www.jedec.org). Among many other things, this is the body that determines standards for physical outlines of IC and discrete device packages.

job shops: see **contract manufacturers**.

known good die (KGD): products such as hybrid and multi-chip modules or chip-on-board devices need die that have been tested as thoroughly as packaged ICs. Presently, however, it is very difficult to subject die (IC chips) to the same degree and quality of tests as packaged ICs, so it is an area of needed improvement to be able to provide die which are known to be as good as packaged die.

land: the copper trace on a printed wiring board, connecting one pad to another or to a via.

large scale integration: originally a term used to describe ICs with transistors numbering between 100 and 1000. This term is not used much any more.

lead frame: the metal frame onto which the IC chip is mounted and to which it is attached electrically. These leads become the I/Os for that IC chip.

lead pitch: the distance between the center points of two adjacent leads on an IC package. For many years, the standard lead pitch was 100 mils. In more recent years, especially for surface-mount technology, the lead pitch has been reduced to 50, 30, 25, and even 20 mils. A finer lead pitch allows many more I/Os per side of the IC package, but at the cost of greatly increased lead fragility and increased printed wiring board assembly costs.

leakage: an ideal capacitor would hold an electrostatic charge indefinitely. Actual capacitors leak, which means that the charge stored in a capacitor will eventually be gone due to leakage. The primary leakage path is generally through the dielectric of the capacitor, but can also be through the capacitor coating material or the circuit to which the capacitor is connected. Capacitor types with very low leakage include most of the plastic types and the specialty types (oil and vacuum).

mandrel: a round shaft, around which is created the component of interest. Carbon and metal film resistors are manufactured on a ceramic mandrel, as are some wirewound resistors.

MCM (multi-chip module): a module with more than one IC chip, but without any thin or thick film resistors.

medium scale integration: a term used in the 60s and 70s to describe ICs with transistors numbering between 10 and 100. This term is not used much any more.

metal electrode face (MELF): a cylindrical SMT package type which looks much like an axial resistor but without the leads. Used for both resistors and diodes.

metal film resistor: a resistor made by depositing a film or ink of metal oxide resistive material on a ceramic mandrel. Because they are as inexpensive as carbon composition resistors but have much better performance in tolerance, temperature drift, and age drift, they have largely displaced carbon composition resistors. (See also **carbon film resistor**)

mixed mount technology: including both surface mount technology (SMT) and through-hole technology on the same PWB. Most PWBs manufactured today are of this type, since it is so difficult to design PWBs with only SMT.

module: the physical package which houses the IC chip, and which provides leads to connect the chip to the module carrier (usually a PWB). It is level 1 of packaging levels.

modulus: see **Young's modulus**.

modulus (of elasticity): see **Young's modulus**.

molecular beam epitaxy: a process used in IC manufacturing for adding materials (mostly metals) to the IC chips. It requires an ultra-high vacuum, and can be precisely controlled.

monochromatic: of only one color. Monochromatic light is highly desirable in IC photolithography because it gives greater resolution than multi-chromatic light. Lasers are a good example of a light source which is highly monochromatic; incandescent light bulbs are a good example of a light source which is highly multi-chromatic.

monocrystalline: a solid with an atomic structure of a single crystal orientation. Monocrystalline silicon is essential to make integrated circuits, and where the silicon has defects in this monocrystalline structure, there will be defects in the operation of the circuit built with it.

motherboard: Another type of level 3 device (see also **backplane**), consisting of board connectors, level 1 components, level 2 components, and a printed wiring board interconnecting them all. Motherboards are common in personal computers, tablet computers, and even cell phones.

multi-chip module: see **MCM**.

multi-layer PWB: a printed wiring board with more than 2 layers, and usually 4 to 6. Multi-layer PWBs typically do not have more than 14 layers. The cost for a multi-layer PWB is proportional to the number of layers, but even a 4-layer PWB is significantly more expensive than a 2-layer PWB.

network resistor: a type of resistor in which there are multiple resistive elements on a common ceramic (usually alumina) substrate. The resistive elements are made from either resistive ink, as in carbon film and metal film resistors, or a thin film of the resistive material. Network resistors provide higher reliability, lower cost, and higher packaging density than the equivalent number of axial resistors.

NRE costs: Non-Recurring Engineering costs; the costs incurred in creating the first working prototype, plus the costs incurred in making the prototype ready for volume manufacturing. NRE costs must be amortized over the volume of sales after manufacturing and sales have begun, or the company cannot recover these expenses. These are basically the costs of design and development.

organic solderability preservative (OSL): a type of coating used over the exposed pads of a PWB to keep oxides from forming and therefore preserve its solderability. Organic means it is made of a type of plastic. This is used in lieu of hot air solder leveling where thickness dimensions are critical.

oxidation: in electronics, this primarily means the formation of an oxide layer on the surface of a metal. Oxygen combines readily with most metals; this resulting oxide is poorly attached to the base metal, is usually non-conductive, and greatly impedes good electrical connections and good solder joints. Metals that do not oxidize readily include the precious metals gold and platinum.

parasitics: essentially all circuit devices exhibit undesirable behavior due to the presence of one or more undesired elements; these are parasitic elements, or simply parasitics. For example, a resistor provides the desired number of Ohms, but it also gives a small but finite amount of capacitance and inductance; this parasitic capacitance and inductance can significantly affect the performance of the resistor at frequencies where these parasitic elements are no longer insignificant.

permeation: the act of passing through a substance or mass. The classic example in electronics is the permeation of water vapor (humidity) through the material of the module to impinge on the IC chip.

permittivity: the degree to which a material permits the establishment of an electrostatic field within it. Materials with a high permittivity love to have electrostatic fields established within them, and do so readily. High permittivity materials are desirable for making capacitors. Among the materials with the lowest permittivity is air, which has a permittivity only slightly higher than that of a perfect vacuum. By definition, a perfect vacuum has a permittivity of 1.00.

PGA: Pin Grid Array. An IC package type in which the I/Os are pins in an array under the chip. A PGA package is a through-hole type of package. Array-type packages, such as the BGA and the PGA, allow many more pins in a given area at a given pitch.

photochemical vapor deposition (PCVD): a process used in IC manufacturing for adding material to the wafer. Chemicals vapors with the desired materials in the gaseous mixture are exposed to a light source, which initiates a chemical reaction. One of the byproducts of this chemical reaction is the material which is to be added to the wafer.

photolithography: the process by which light is used to make patterns on IC wafers in the process of IC manufacturing. Photolithography is an exacting process, which can fail in many ways; it is also a major key to increasing the resolution of an IC manufacturing process. A great deal of research has gone and is going into this process.

photoresist: a light-sensitive material which protects the areas that are not to be affected by a given process. A photoresist material has three stages: 1) the liquid stage, in which the photoresist is spread over all the area to be photo-defined; 2) the partially-hardened or partially-softened stage, which is a solid stage; 3) the fully-hardened stage, which is the stage it is in when it protects the desired areas. In a positive photoresist, light acts to soften the material, allowing it to be dissolved readily. In a negative photoresist, lights acts to fully harden the material.

pick-and-place robot: a specialized type of robot used in the assembly of components onto PWBs. Their primary function is to pick up the appropriate component from the corresponding component feeder, then place that component accurately and quickly onto the PWB. Pick-and-place robots are used extensively in the electronics manufacturing industry.

pitch: the distance between the centers of I/O pins along a given axis. The classic example is the DIP, which has its pins on a pitch of 100 mils.

polycrystalline: having many (poly) crystal sizes and orientations. Many solid materials fall into this classification, including steel, copper, aluminum, and most other ordinary metals.

potentiometer: a three-terminal variable resistor in which the outer two terminals are connected to a voltage difference and the middle terminal, termed the wiper, provides a potential (voltage) which is proportional to the movement of the physical adjustment of the potentiometer.

pre-impregnated: printed wiring boards are commonly made up of layers of woven fiberglass which has been immersed in liquid epoxy and then partially cured. These sheets of composite material have been *pre-impregnated* with epoxy. See also **prepreg**

prepreg: a composite of the woven fiber material, pre-impregnated (soaked) with the binder material (usually epoxy), and partially cured. Prepreg material is made in sheets, many of which are laid up together to form the insulative layer of a PWB.

printed circuit board: also known as printed wiring board (PWB). A PWB is the substrate that provides the circuit for connections for electronic components. It is the most common method of electrically connecting and mechanically attaching electronic and electrical components for use in an electronic product. Although actually made by an etching and not a printing process, the term *printed* has survived.

process window: In manufacturing an electronic device or product, we would ideally be able to set all the relevant parameters for production and they would remain at that set level. However, the variation common in this world causes that these parameters vary as a function of many other parameters, sometimes in complex interactions. Thus, to successfully manufacture devices and products, it is necessary to determine the minimum and maximum levels for each parameter over which the process can continue to successfully function. These minimum and maximum levels are the process window. For example, if the heating process for making carbon composition resistors calls for the mold to be held at 240°C ±2°C, for a period of 15 seconds ±1 second, the process windows would be ±2°C and ±1 second.

PWB: Printed Wiring Board; also often referred to as a Printed Circuit Board, or PCB. This is the carrier onto which are fastened the discrete components and the ICs necessary to complete a circuit.

quality: a measure of the consistency and overall process improvement orientation of a product or a company. A product of high quality is one which consistently meets or exceeds its performance criteria, and which is continually being improved.

radial: having leads coming out the radius of a cylindrical object. Components with radial leads include some capacitors, TO-5 transistors, and some switches.

reflow of through-hole (ROT): the practice of screen-printing solder paste on the pads of through-hole components, followed by insertion of the through-hole components into their holes and reflow of the assembled board. This practice allows both through-hole and surface-mount components to be soldered in a single process, instead of having to wave solder the through-hole components, followed by reflow soldering of the surface-mount components.

resin substrate materials: the various materials used for the resin of the composite materials commonly used for making PWBs. These resin materials include epoxy, polyimide, and phenolic polymers. Most PWBs are made from a composite of both resin and fiber.

reticle: a mask covering only a portion of the entire wafer. A reticle must be stepped and repeated many times to expose an entire wafer.

rheological: having to do with the flow properties of a material. Rheological properties are important in solder paste, so that it will properly flow through the openings in the screen through which it is forced.

rheostat: a two-terminal variable resistor in which the outer two terminals are connected in series with a voltage, and which provides a current which is proportional to the movement of the physical adjustment of the potentiometer. *Rheo* is from the Greek word for flow; rheostats are used to adjust the flow of electricity (current), and are generally much larger than their close relative, the potentiometer.

screening: the process of testing finished devices or products for the purpose of determining which are functional and which are not. Screening is one of three main purposes of testing, the other two being process monitoring and improvement and fault isolation.

self-resonance: resonance is a zero-reactance condition that occurs at a specific frequency for a given combination of inductance and capacitance. Since all electronic components contain parasitic elements, an inductor will have some capacitance, a capacitor will have some inductance, and a resistor will have some inductance and some capacitance. Due to the presence of these parasitic elements, these components will resonate with themselves at a certain frequency. This is usually an undesirable condition.

single in-line memory module (SIMM): the most common form factor for memory in personal computers. A SIMM is a small PWB (usually about 2 cm high by 12 cm long) with memory modules on both sides.

slump: to decrease in the vertical dimension by spreading out in the horizontal dimension. After solder paste has been screen printed onto the pads of a circuit board, it would ideally stay in exactly the shape in which it was deposited. Under normal conditions, however, the solder paste will eventually slump, which can cause it to spread to adjacent pads. This is not desirable, as it may cause a short when reflow soldering occurs.

small scale integration: a term used in the 60s and 70s to describe ICs with transistors numbering between 2 and 10. This term is not used much any more.

"smart" cards: cards that look mostly like an ordinary credit card, but which have on-board intelligence (a microcontroller) and storage (memory). Smart cards have not caught on well in the United States, but are widely used in the EU and elsewhere.

solder: an alloy of two or more metals, the combination of which has a lower melting point than that of any of the constituent metals. In electronics, the most common type of solder used is a mixture tin (Sn), silver (Ag) and copper (Cu) at a ratio of 96.5%/3%.0.5%, and is known as SAC 305.

static electricity: the form of electricity characterized by a buildup of charge. Static electricity is most often noticed when a discharge of it occurs, in the form of a small momentary arc.

stepping: a term used in photolithography of IC chips, referring to the process of exposing a portion of the wafer to the light source through the reticle, then moving the wafer or the reticle to another position and repeating the process. Each step is one repetition of this cycle. The stepping process is repeated until all the wafer has been exposed.

stepper: the primary photolithographic tool used in modern IC manufacturing. A stepper exposes a portion of the wafer through a reticle, thus photodefining the areas on the IC. After each portion of the wafer is exposed, the stepper precisely steps the wafer to a new area and repeats the exposure process, repeating until the entire wafer is thus exposed.

stripping: the chemical process of removing hardened photoresist from a wafer or PWB during the manufacturing process. Stripping is usually a wet chemical process; after exposure to a stripper chemical, the hardened photoresist lifts off the substrate.

surface mount technology (SMT): a term which refers to any technology which enables a device to be mounted on the surface of a PWB, as opposed to being mounted in a through-hole. SMT enables much higher component density because the leads are on a smaller pitch and the components may be mounted on both sides of the PWB.

swage: the second step in the three-step process of mating the two halves of a connector. The first step is *contact*; the third step is *swipe*. Swage is the step in which at least one of the two electrically conducting surfaces deforms to provide a spring force to keep the surfaces together.

swipe: the third step in the three-step process of mating the two halves of a connector. The first two steps are *contact* and *swage*. Swipe is the step in which the two electrically conducting surfaces rub against each other, for the purpose of removing any built-up dust or oxide.

temperature coefficient: coefficient means multiplier; temperature refers to the fact that many parameters vary as a function of temperature. This book addresses some examples such as *coefficient of thermal expansion (CTE)* and *temperature coefficient of resistance*.

temperature coefficient of resistance: the resistance of most electronic devices changes as a function of temperature. For example, the resistance of metals increases with temperature; the resistance of certain types of semiconductors decreases with temperature. The temperature coefficient of resistance must be considered in the design of many electronic products, as it may cause the resistances in the product to vary more than originally anticipated.

temperature cycle: the sequence of going from powered-off (and therefore at room temperature) to powered-on (and therefore at some more elevated temperature due to the power dissipated by the electronic devices), and back to powered off. Each temperature cycle of an electronic device or product creates one cycle of stress and strain due to the thermomechanical forces of the expansion and contraction of the materials from which the devices or products are made.

temperature drift: the variation of a given parameter as a function of temperature. This term is synonymous with *temperature coefficient*.

temperature stability: the degree to which a parameter does not vary as a function of temperature. It is usually very desirable to be able to design with devices characterized by parameters which do not vary with temperature; such designs exhibit a great degree of temperature stability.

thermal conductivity: the intrinsic ability of a material to conduct heat, measured as a function of the surface area divided by the length (cm^2/cm, or just cm). It is the inverse of thermal resistivity, and the units are W/°C.

thermal resistance: opposition to the flow of heat, and the inverse of thermal conductance. Materials with high thermal resistance are known as thermal insulators, and include examples such as styrofoam and fiberglass. Materials with low thermal resistance (high thermal conductance) include examples such as copper and aluminum. The units are °C/W.

thermomechanical: having to do with the mechanical properties and considerations that must be studied when considering the effects of heat. One of the primary thermomechanical considerations is the different CTEs of the materials which make up electronic devices.

through-hole in solder paste: see **reflow of through-hole (ROT)**.

through-hole mount technology: any technology which requires or enhances the use of through-hole components. This is the counterpart to *surface-mount technology*.

time of flight: the time a change in voltage takes to propagate from one point to another. Time of flight has its absolute minimum limited by the speed of light, and is proportional to the resistance of the conductive material and the dielectric constant of the insulating material.

tolerance: the total amount of variation of a parameter. For instance, if one obtains 1,000 resistors specified at 100 Ohms each, and in actuality they range in value from 95 Ohms to 105 Ohms, their tolerance would be ±5%.

transistor outline: the physical dimensions of a transistor package type. These are preceded by the prefix TO-, and the various transistor outlines are standardized by JEDEC. (See also **diode outline**)

underfill: the material between a flip-chip and the substrate onto which it is electrically and physically fastened. Used primarily to help spread the substantial thermomechanical forces extant when a chip is flip-chip mounted to an organic substrate, because of the significant difference between the CTE of the chip and the CTE of the organic substrate. Underfill material is usually an elastomer which hardens after curing; the elastomeric properties of the hardened underfill material help substantially to absorb these thermomechanical expansion and contraction forces.

ultra large scale integration: originally a term used to describe ICs with transistors numbering more than 10,000. This term is not used much any more.

value range: the minimum and maximum values available in a given type of device. Ideally, a component would be available in any desired value. In practicality, there is a range of about 6 orders of magnitude for most resistor types and most capacitor types.

very large scale integration: originally a term used to describe ICs with transistors numbering more than 1,000. This term now is used generically to describe the state of the art in IC manufacturing, which is well over 100 billion transistors per IC.

via: a vertical connection between wiring layers in an IC chip or a PWB.

vitreous: literally, glass-like. There are many materials in the vitreous family, including ceramics, porcelain, china, and of course, glass. Vitreous materials are characterized by being very hard and usually very brittle.

voltage difference: the maximum difference between the voltage potentials on a device or PWB. For example, if a PWB is powered by +5 Volts and:15 Volts, the voltage differential is 20 Volts.

voltage rating: the maximum voltage for which a device is rated for operation. The voltage rating for a device should not be exceeded, since that will shorten its lifetime.

volumetric efficiency: packing the maximum amount of a desired property in the minimum amount of volume. This term is particularly used in describing capacitors; those with the highest volumetric efficiency are aluminum and tantalum electrolytic capacitors, which give the highest amount of capacitance possible for a given volume.

wafer: the thin slice of semiconducting material (usually silicon) on which IC chips are fabricated. Silicon wafers typically range in thickness from 12 to 20 mils, and in diameter from 5" to 12" (125 mm to 300 mm).

wave soldering: this is the process of passing an assembled printed wiring board over a wave of molten solder, which causes all the exposed metal to take on solder. By design, only the points (connections) where solder is desired are exposed during wave soldering. Also sometimes termed flow soldering.

wet (chemical) etching: In the process of making integrated circuits, it is often necessary to remove one material from designated areas of the chips. Wet or chemical etching does this by exposing the unprotected areas of the chip to a chemical solution, which reacts with the unprotected areas of the chip and etches them away. Although wet etching is less expensive than dry etching, it suffers from the disadvantage of being an isotropic process, which does not permit the higher resolution provided by dry etching, which is an anisotropic process. (See also **dry etching**). Wet etching is still widely used in PWB manufacturing.

wire-wound resistor: a resistor that is made of a coil of resistive wire (typically nickel-chromium or nichrome). Wire wound resistors are specialized resistors of two types: precision or power.

wirebonding: the process of electrically connecting each pad of an IC chip to each lead of the leadframe. Wirebonding is done one bond at a time, either by thermocompression (usually for gold wires), by ultrasonic (usually for aluminum wires), or by thermosonic (a combination of the two).

wire-wound potentiometer: a potentiometer that is made of a coil of resistive wire (typically nickel-chromium or nichrome). Wire-wound potentiometers are designed for use primarily as rheostats.

wire-wound rheostat: see **wire-wound potentiometer**.

yield strength: the point in the stress-strain curve at which the material fails (breaks). Materials with a high yield strength are desirable in electronic packaging, because they can better withstand the thermomechanical forces to which electronic devices are subjected.

Young's modulus: the ratio of tensile stress to tensile strain for a given material. This is often referred to as the *modulus of elasticity* or simply as the *modulus*. Materials with a high modulus are stiff, such as ceramic or glass; materials with a low modulus are flexible, such as rubber.

INDEX

D

H

I

J

K

L

M

P

Q

R

T

W

Y